Recent Results in Cancer Research 150

Springer
Berlin
Heidelberg
New York
Barcelona
Budapest
Hong Kong
London
Milan
Paris
Santa Clara
Singapore
Tokyo

R. Engenhart-Cabillic A. Wambersie (Eds.)

Fast Neutrons and High-LET Particles in Cancer Therapy

With 83 Figures and 36 Tables

Springer

Prof. Dr. Rita Engenhart-Cabillic
Universität Marburg, Medizinisches Zentrum für Radiologie
Abteilung für Strahlentherapie, Baldinger Straße
D-35033 Marburg

Prof. Dr. André Wambersie
Université Catholique de Louvain, Cliniques Universitaires St.-Luc,
Avenue Hippocrate 10, B-1200 Bruxelles

ISBN-13: 978-3-642-78776-8

ISSN 0080-0015

Library of Congress Cataloging-in-Publication Data
Fast neutrons and high-LET particles in cancer therapy / R. Engenhart, A. Wambersie (eds.). (Recent results in cancer research; 150) Includes bibliographical references and index. ISBN-13: 978-3-642-78776-8 e-ISBN-13: 978-3-642-78774-4
DOI: 10.1007/978-3-642-78774-4 1. Fast neutrons–Therapeutic use. 2. Cancer–Radiotherapy. I. Engenhart, R. (Rita), 1953– . II. Wambersie, A. III. Series. [DNLM: 1. Neoplasms–radiotherapy. 2. Fast Neutrons–therapeutic use. 3. Radiotherapy, High-Energy–methods. W1 RE106P, v. 150 1998 / QZ 269 F251 1998] RC261.R35 vol. 150 [RC271.N48] 616.99'4 s–dc21 [616.99'40642] DNLM/DLC

Production: PRO EDIT GmbH, Heidelberg
Typesetting: K+V Fotosatz GmbH, Beerfelden

SPIN 10127888 19/3133-5 4 3 2 1 0 – Printed on acid-free paper

Preface

The initial pioneer work in fast neutron therapy was performed by Stone at Berkeley in 1938. A fast neutron therapy programme was initiated in 1970 at the Hammersmith Hospital in London, and a few years later in several centres in Europe and in the United States. After almost 30 years of widespread experience, it is timely and useful to evaluate the results of neutron therapy in a comprehensive and objective way. That is the aim of this book.

In Europe, 14 centres have applied fast neutrons in cancer therapy, using different types of generators:

- "Low-energy" cyclotrons able to produce d(13–16)+Be neutron beams: Hammersmith and Edinburgh in the UK; Essen and Dresden in Germany; Krakow in Poland.
- (d+T) generators: Hamburg, Heidelberg, and Münster in Germany; Amsterdam in The Netherlands; Glasgow in the UK.
- "Higher-energy" cyclotrons: Louvain-la-Neuve in Belgium (p(65)+Be); Orléans in France (p(34)+Be); Clatterbridge in the UK (p(62)+Be); and, since 1997, Nice in France (p(50)+Be).

Some centres completed large clinical programmes from which important conclusions could be derived about the potential benefit of high-LET radiations. In other centres, the facility was shut down quite rapidly, in some cases because of patient recruitment problems but in most cases because of technical difficulties. In particular, all (d+T) facilities are now closed. Indeed, the low dose rates available with the (d+T) generators have always caused serious limitations in the selection of the treatment modalities.

In comparing the relative merits of fast neutron therapy and photon therapy, it has been recognised that fast neutron therapy has often been applied in suboptimal technical conditions. Poor beam penetration and collimation were compensated for in some institutions by sophisticated treatment planning and patient positioning for rather superficial tumours; for deep-seated tumours, however, they were definitely a handicap.

The situation has significantly improved during the past 15 years, in Europe and worldwide. At some facilities fast neutrons can be applied in the same advanced technical conditions and with the same physical selectivity as photons, as far as beam penetration, collimation and isocentric gantry are concerned (for example, Seattle and Detroit in the USA and NAC-Faure in South Africa).

The rationale for introducing fast neutrons (or, more generally, high-LET radiations) in cancer therapy in the 1970s was based on radiobiological considerations:

- Originally, reduction in the OER with increasing LET
- Reduction of differences in radiosensitivity related to the position of the cells in the mitotic cycle
- Furthermore, reduction of the relevance of the repair phenomena.

These arguments remain fully valid and have not been contradicted by more recent radiobiological findings. However, they imply the need for the development of "predictive tests" allowing the radiation oncologist to select the patients who could benefit most from high-LET radiations. This remains today an important research issue.

Most of the efforts in the field of hadron therapy in Europe were initiated and coordinated by the European Organization for Research and Treatment of Cancer (EORTC)-Fast Neutron Therapy Group. This group, founded in 1972 by the late Professor Klaas Breur from Amsterdam, became the European Hadron Therapy Group in 1995 (hadrons include fast neutrons, protons and heavy ions, i.e. non-conventional therapy beams). The aim of the group was to explore the possibilities of non-conventional radiations in cancer management and ultimately to initiate randomised clinical trials in order to be able to demonstrate the benefit of the new types of beams for some tumour types or sites.

An important achievement of the European Hadron Therapy Group was to initiate and coordinate intercomparisons at the different facilities, embracing dosimetry, microdosimetry and radiobiology.

As far as dosimetry is concerned, intercomparisons were performed at all the European neutron therapy facilities. As indicated in the first chapter of this book, an initial European protocol was established and applied. Because of significant discrepancies with the US AAPM protocol, efforts towards a common protocol were initiated under the auspices of the International Commission on Radiation Units and Measurements (ICRU) and an agreement was reached and accepted worldwide (see ICRU Report 45, 1989, "Clinical neutron dosimetry. I. Determination of absorbed dose in a patient treated by external beams of fast neutrons").

Intercomparisons aiming at specifying radiation quality at the different neutron facilities are dealt with in the second chapter. Radiation quality can be specified in terms of microdosimetric spectra and/or in terms of RBE values observed for well-defined biological systems. It is important to stress that both microdosimetric and radiobiological intercomparisons were performed by the same teams of biologists and experts in microdosimetry visiting the different facilities.

The side effects after neutron and high-LET therapy have always been a cause of concern. The complications observed at the level of the "treated volumes" and of the "irradiated volumes" (see definitions of these concepts in ICRU Report 50, 1993) are discussed for the different tumour localisations in the fourth to ninth chapter, and especially in the tenth chapter. However, of particular concern is cancer induction after neutron and high-LET therapy. No human data are presently available, and the best estimates derived from the radiobiological data are discussed in the third chapter.

The fourth to ninth chapters contain an extensive review of the published clinical data. There is emphasis on the data from the European centres; however, relevant data obtained worldwide are included. Neutrons were shown to be superior to conventional radiations for the treatment of some tumour types or sites, e.g. salivary gland tumours and prostatic adenocarcinomas. The most recent and complete data on this are presented. For other tumour types or sites, such as soft tissue sarcomas and chordomas, or for palliative treatment of inoperable/recurrent rectal carcinomas, the available neutron data are analysed and the clinical indications are discussed.

For some tumour localisations, data from different centres, in particular from Europe, have been pooled, analysed, and presented by a given group of authors. It is believed that this approach improves the quality and objectivity of the information.

The complications resulting from the fact that the first neutron treatments were often applied in suboptimal physical conditions have been mentioned above. In that respect, the logical trend in application of high-LET radiations in better physical conditions is to move towards heavy-ion beams, such as carbon or neon ions. This trend is equivalent to that observed with low-LET radiations, where proton beams have began to complement photon beams.

The first heavy-ion therapy programme was initiated in Berkeley in 1977 and ran until 1992. Although there were some limitations in patient recruitment and in machine availability, important information was obtained about the potential value of heavy ions from the series of 433 patients who could be treated. At the time when, unfortunately, the Berkeley facility was being closed, a new carbon-ion therapy programme was started in Chiba, Japan. There is no limitation concerning the machine time, and three treatment rooms are planned. From the clinical point of view, the major efforts were focused on the treatment of the most frequent tumours in Japan, i.e. those of oesophagus, bronchus, and liver. The first patients were treated in 1994, and a total of 300 patients were treated by the end of 1997.

In Europe, a heavy-ion therapy programme was initiated at GSI-Darmstadt, in Germany, jointly with the German Cancer Research Centre (DKFZ) and the Department of Radiation Oncology of Heidelberg University. The clinical programme is orientated differently compared to the Chiba programme. There are strict limitations on machine time, but full advantage is taken of the scanning beam system and the possibility of modulating the carbon-ion beam energy. The selected clinical indications are thus, logically, radioresistant tumours that are irregular in shape and located close to critical normal structures for which there is at present no reasonable therapeutic alternative (e.g., some tumours of the skull base). For these types of localisations, the characteristics of the beam at GSI can be fully exploited.

The first two patients, with target volumes located at the base of the skull, were treated at the end of 1997. A chapter on heavy ions thus has a definite place in this book and helps to make it up to date.

The cost of the treatment facility will, for the moment, limit the development of heavy-ion therapy programmes. However, as has been observed with neutrons and protons, it can be expected that new technical developments will make the cost of heavy-ion therapy more affordable. It is recognised, however, that the investment, at present, is of another order of magnitude compared to that necessary for neutron or proton generators.

This book on high-LET radiations would not have been complete without a review of the present status of boron neutron capture therapy (BNCT) and a discussion of its future perspectives. BNCT is a special type of high-LET radiation therapy, with the intention of achieving selectivity at the cellular level.

BNCT was initiated at MIT (Boston) and at the Brookhaven National Laboratory (BNL) in the 1950s. It was introduced in Japan by Hatanaka in 1968 for the treatment of brain tumours, and more than 130 patients were treated, including at least 12 American patients. The results reported from Japan were severely questioned by the radiotherapy community, in particular in the USA.

However, the rationale for BNCT, and in particular its attempt to achieve selective irradiation at the cellular level, is attractive and could open new perspectives in radiation therapy, especially for some tumours for which no efficient treatment is available at the moment (such as glioblastomas).

For these reasons, the US Department of Energy has supported BNCT research programmes and clinical application programmes at MIT and at BNL. A total number of 40 patients with glioblastomas were treated between September 1994 and December 1997. In Europe, the European Commission has supported a BNCT project in Petten, The Netherlands. The first treatment was performed in October 1997, and three patients had been treated by the end of 1997.

Combination of BNCT with fast neutron beam therapy has also been proposed in order to enhance the selectivity of fast neutron therapy. This combination has indeed been used in some neutron therapy centres, such as Essen in Germany and Seattle in the USA.

The two closing chapters thus appear timely, making this book on cancer therapy with high-LET radiations as complete and up to date as possible, especially regarding the situation in Europe.

Rita Engenhart-Cabillic
André Wambersie

Contents

List of Contributors*

Auberger, T.[88, 100, 137]
Austin-Seymour, M.[125]
Barth, R.[183]
Blue, T.[183]
Brackrock, S.[88]
Breteau, N.[88, 113, 148]
Cho, P.[125]
Cohen, L.[156]
Debus, J.[113, 170]
Demasure, M.[148]
Engenhart-Cabillic, R.[88, 100, 113]
Engels, H.[54]
Favre, A.[88, 100]
Gahbauer, R.[183]
Goodman, J.[183]
Grecula, J.[183]
Griffin, T.W.[125]
Guculette, J.[31]
Gupta, N.[183]
Höver, K.H. 113
Huber, P.[88]
Kraft, G.[170]
Koh, W.-J.[125]
Koppe, H.[88]
Krüll, A.[88, 100, 113]
Jäckel, O.[170]
Laramore, G.E.[125]
Lescrainier, J.[148]
Lessel, A.[88, 100]
Lindsley, K.L.[125]
Menzel, H.G.[31]
Michenet, P.[148]
Pihet, P.[31]
Pötter, R.[113]
Prott, F.J.[88, 100, 113]
Reuschel, W.[137]
Russell, K.J.[125]
Sabattier, R.[148]
Sauerwein, W.[183]
Schmidt, R.[1, 100]
Schwarz, R.[88, 100]
Soloway, H.[183]
Stelzer, K.J.[125]
Vynckier, S.[1]
Wambersie, A.[31, 54, 183]
Wannenmacher, M.[170]

* The address of the principal author is given on the first page of each contribution.

[1] Page on which contribution begins.

The Physical Basis for Radiotherapy with Neutrons

S. Vynckier[1] and R. Schmidt[2]

[1] Cliniques Universitaires St-Luc, Avenue Hippocrate 10, Brussels, Belgium
[2] Universitäts-Krankenhaus Eppendorf, Martinistrasse 52,
D-20246, Hamburg, Germany

Abstract

Radiotherapy with neutrons requires a large quantity of physical information about neutrons and their interaction with matter. Primarily the absorbed dose at a reference point needs to be determined. For treatment planning the dose distribution in a phantom must be measured and algorithms for the simulation of the dose distribution in a patient must be available. As neutrons interact with matter in a more complicated way as high-energy photons and electrons commonly used in radiotherapy, biological effects based on microdosimetric data are used for treatment planning. This paper presents a brief summary of the neutron sources used in radiotherapy. The dosimetry of the clinical neutron beams is described. Special aspects of the treatment planning with fast neutrons are discussed. For further radiobiological interpretation the fundamentals of microdosimetry are described. Finally recent and future developments in the field of physics for neutron therapy are mentioned.

Neutron Sources for Radiotherapy

Physical Properties of Neutrons

The neutron is an uncharged nuclear particle and therefore cannot be accelerated or deflected. Because of its physical properties its interaction with matter is based on reactions with the nuclei of the matter, mainly by elastic or inelastic scattering or in the low-energy region by capture processes.

Neutrons are generally denominated according to their kinetic energy E_n:

Thermal neutrons	$E_n < 0.5$ eV
Intermediate neutrons	0.5 eV $< E_n <$ 10 keV
Fast neutrons	10 keV $< E_n <$ 20 MeV
Relativistic neutrons	$E_n >$ 20 MeV

Recent Results in Cancer Research, Vol. 150

Three principally different neutron sources can be used in radiotherapy: fission and radioactive sources, nuclear reactors and accelerators.

Production of Neutron Beams for Radiotherapy

Nuclear Reactor

Neutrons produced in a nuclear reactor cover an energy range from thermal to fast neutrons (Fig. 1). Converters and filters can be used to absorb thermal and intermediate neutrons, thus hardening the beam so that a mean energy of 1–2 MeV can be otained. Neutrons from reactor sources are used for the treatment of superficial tumours or for neutron capture therapy. In boron neutron capture therapy (BNCT) a ^{10}B carrier agent is used to concentrate the boron in the tumour. The irradiation of ^{10}B by thermal neutrons reveals high-linear energy transfer (LET) α-particles via the reaction $^{10}B(n, \alpha)\,^{7}Li$. The neutrons from a nuclear reactor or the thermalized neutrons from a fast neutron beam can be used. In the latter case the BNCT can be handled as a concomitant boost to fast neutron therapy, whereas for BNCT with reactor neutrons the energy deposited by the α-particles is the therapeutic basis. The mean free path length of the neutrons can be increased by appropriate filtering to obtain penetration into deeper sites.

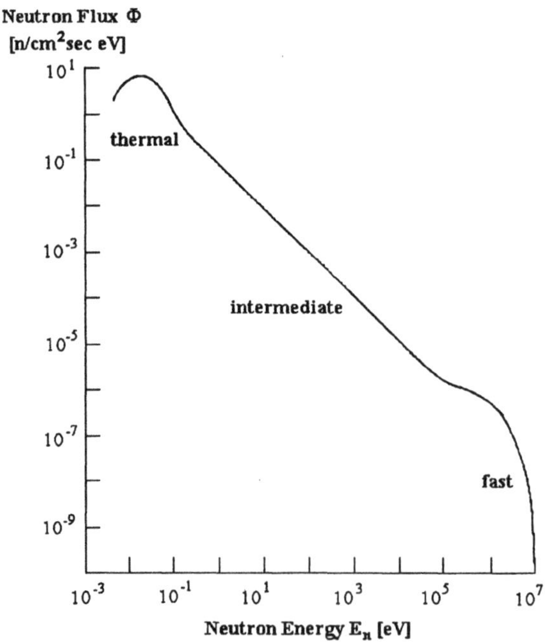

Fig. 1. Typical reactor neutron spectrum

Radioactive Sources

The most common neutron-emitting isotope is ^{252}Cf. The emitted neutrons have a mean energy of 2 MeV; the radioisotope has a half-life of 2.65 years. ^{252}Cf sources can be used in remote afterloading systems for interstitial and intracavitary therapy in a way comparable to gamma sources.

A second type of "radioactive" neutron sources uses the reactions $^{9}Be(\alpha, n)^{12}C$ or $^{9}Be(\gamma, n)\,2\alpha$. The α- or γ-particles required for these reactions are obtained from radioactive sources. As α-emitting source ^{238}Pu and ^{241}Am radioisotopes are used, as γ-emitter ^{226}Ra. These sources are used not for radiotherapy, but for calibration purposes.

Neutron Production by Accelerated Particles

Fast neutron beams for radiotherapy are generated by accelerating protons or deuterons onto a target. Depending on the physical interaction process these reactions are either called fusion or stripping reactions. The processes with importance in radiotherapy are listed in Table 1.

The fusion reactions needs only slightly accelerated particles to enter the modest potentials. The DD reaction has a Q-value of 3.27 MeV. If neutrons with higher energies are required, the incident deuterons have to be accelerated to higher energies. But then a second reaction, called deuteron break-up [$^{2}H(d, np)^{2}H$–2.2 MeV], may occur, leading to lower energy neutrons in the spectrum. High-pressure gas targets mounted to a small cyclotron were temporarily used for radiation therapy. The applied energy distribution was separated into the two energy groups of the neutrons produced by both reactions, as can be seen from Fig. 2 (Höver et al. 1975; Schraube et al. 1975; Watermann et al. 1978).

The high Q-value of the DT reaction provides fast mono-energetic neutrons with a penetration in tissue almost equivalent to ^{60}Co photons. DT generators can be constructed in a compact way so that an isocentric design is possible. Two different generator designs are used for neutron therapy: sealed tubes and rotating targets.

In sealed tubes a mixture of deuterons and tritons is ionized by internal electrodes and accelerated onto an integrated target. The design of a system used at Heidelberg and Münster is given in Fig. 3. The tritium is regenerated so that the sealed tube can be used for several hundreds of beam-hours. Then the whole tube has to be replaced. The neutron tube housed inside a collimator is mounted onto an isocentric gantry. Different circular or rectan-

Table 1. Neutron production by accelearted charged particles

Fusion reaction	Stripping reaction
$^{2}H(d, n)\,^{3}He$ (+3.27 MeV) DD reaction	$^{9}Be(d, n)\,^{10}Be$ (+4.35 Mev)
$^{3}H(d, n)\,^{4}He$ (+17.58 MeV) DT reaction	$^{9}Be(p, n)\,^{9}Be$ (−1.85 MeV)

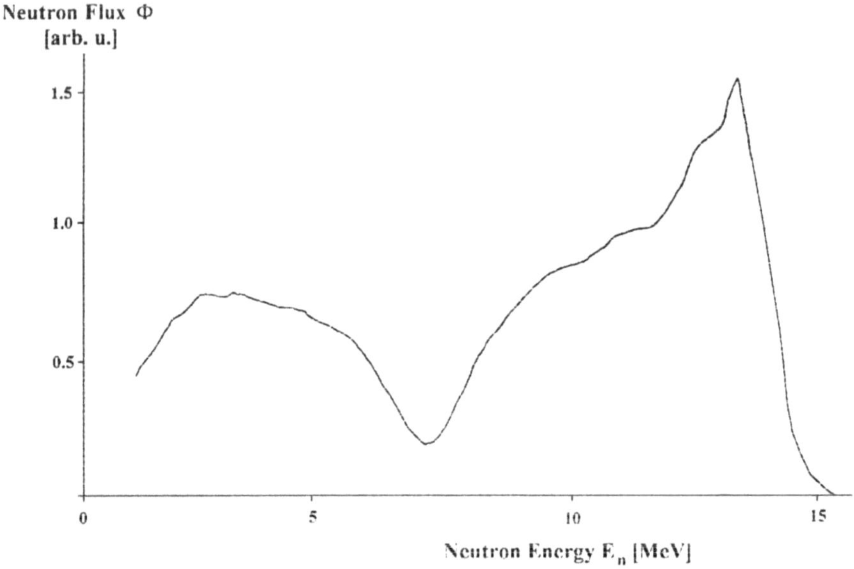

Fig. 2. Neutron energy spectrum of a DD reaction, measured at the high-pressure deuterium target at Heidelberg

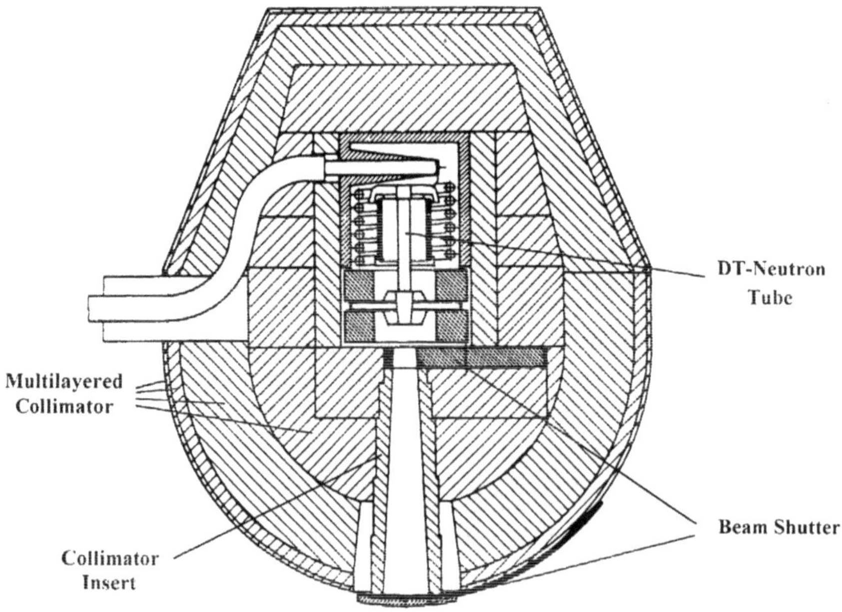

Fig. 3. DT neutron generator KARIN at the German Cancer Research Centre (DKFZ) at Heidelberg

gular field sizes are available by appropriate inserts made out of wood or steel (Höver et al. 1975).

In a rotating target design copper cylinders or discs coated with tritium are used. The deuterons are accelerated by a small linear accelerator onto the rotating target. To extend the half-life of a target to more than 100 beam-hours the deuterons can be separated into their atomic and molecular components. For replacement of the used target by a newly coated copper cylinder or disc, the target assembly is dismounted. Figure 4 shows a design as it is in use at Hamburg. The target and the collimator are housed in an isocentric unit where the treatment couch is rotated around the source. As described above many treatment fields are available by use of collimator inserts (Franke et al. 1983).

Cyclotron-Produced Neutron Beams

Cyclotrons are used for neutron production by accelerating particles onto a beryllium target as explained previously. Two nuclear reactions are applied: the deuteron on beryllium reaction and the proton on beryllium reaction.

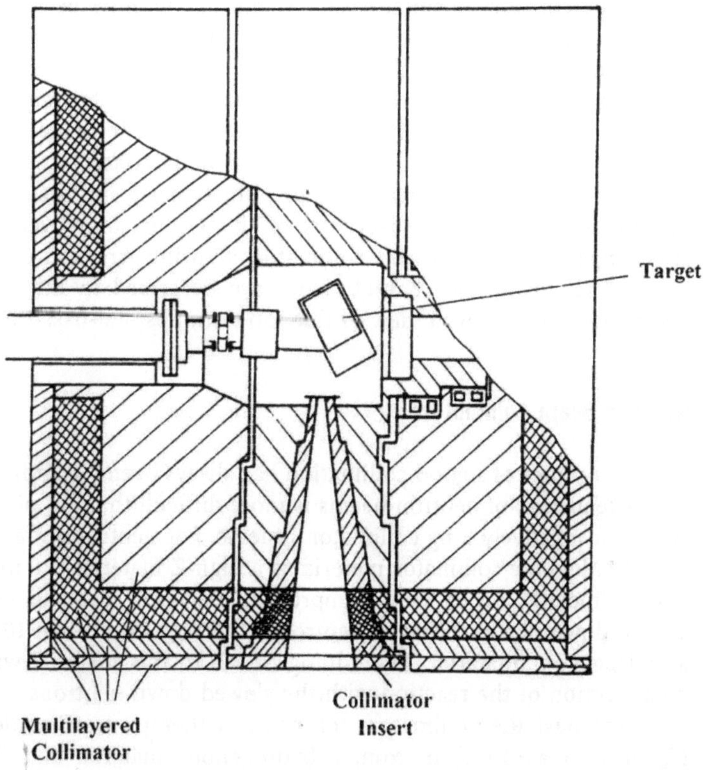

Fig. 4. DT generator at the University Hospital Eppendorf in Hamburg

Table 2. Neutron therapy centres using the d+Be or the p-Be reaction

Centre	Acceleration	Nuclear reaction	Collimator	Gantry
Louvain, Belgium	Cyclotron, variable energy	p(65)+Be	Multileaf	Vertical and horizontal
Fermilab, USA	Linac	p(66)+Be	Inserts	Horizontal
Detroit, USA	Superconducting cyclotron	d(50)+Be	Multirod	Isocentric
Seattle, USA	Cyclotron	p(50)+BE	Multileaf	Isocentric
Faure, South Africa	Cyclotron	p(66)+BE	Variable+leaf inserts	Isocentric
Nice, France	Cyclotron	p(65)+BE	Multileaf	Vertical
Orleans, France	Cyclotron	p(34)+Be	Inserts	Vertical

The latter will produce a larger amount of lower energy neutrons. Therefore, for the protons on beryllium reaction the beam is generally hardened with the insertion of a hydrogenous filter (i.e. polyethylene) (Vynckier et al. 1982, 1983). At present only seven centres are using neutron therapy on a routine basis with the acceleration of particles. The centres and the characteristics of the neutron beams are shown in Table 2.

A more compact hospital-based system has recently been designed by use of superconducting magnets (Maughan et al. 1994). This isochronous superconducting cyclotron, accelerating deuterons to 48.5 MeV, is small enough to be mounted between two large gantry rings. A counterweight acts as beam stop. The system can easily be rotated by 360°. The deuteron beam is incident on an internal beryllium target which it strikes at an angle of 20.3°. The maximum beam current is 15 μA, but typically this centre uses 12.5 μA, yielding a dose rate at the isocentre of 40 cGy/min. Moreover, the installation is equipped with a multirod collimator, allowing irregular-shaped treatment fields. The dosimetric characteristics are described in the next paragraph. Figure 5 shows a typical view of this modern, hospital-based installation.

Neutron Beam Collimators

The importance of a good collimation has always been recognized in radiotherapy. Attenuation of neutron beams is more difficult than for photon beams, resulting in relatively long collimator systems. For neutrons of an energy higher than 10 MeV, the collimator material is a high-Z material like iron or tungsten. Iron is mostly chosen as a compromise between cost and efficiency. Once slowed down, the absorber of neutrons of energy lower than 10 MeV is mostly a hydrogenous material. The adding of ^{10}B to this material will increase the cross section of the reaction with the slowed down neutrons.

In the past the collimators for neutron therapy were made of a mixture (Richard et al. 1981) of iron, a hydrogenous material and ^{10}B. Sometimes also high-density wood was used. The collimation was fixed and could only

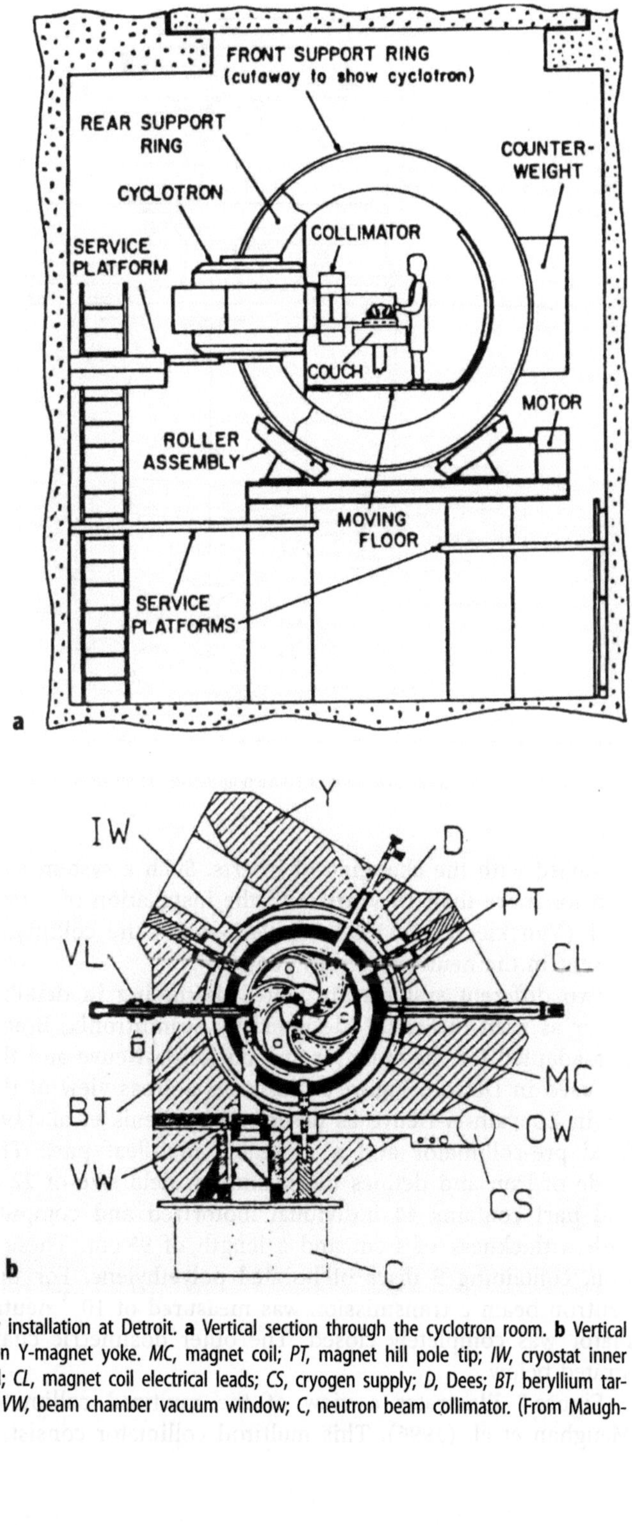

Fig. 5 a, b. Neutron therapy installation at Detroit. **a** Vertical section through the cyclotron room. **b** Vertical section through the cyclotron Y-magnet yoke. *MC*, magnet coil; *PT*, magnet hill pole tip; *IW*, cryostat inner wall; *OW*, cryostat outer wall; *CL*, magnet coil electrical leads; *CS*, cryogen supply; *D*, Dees; *BT*, beryllium target; *VL*, target vacuum lock; *VW*, beam chamber vacuum window; *C*, neutron beam collimator. (From Maughan et al. 1994)

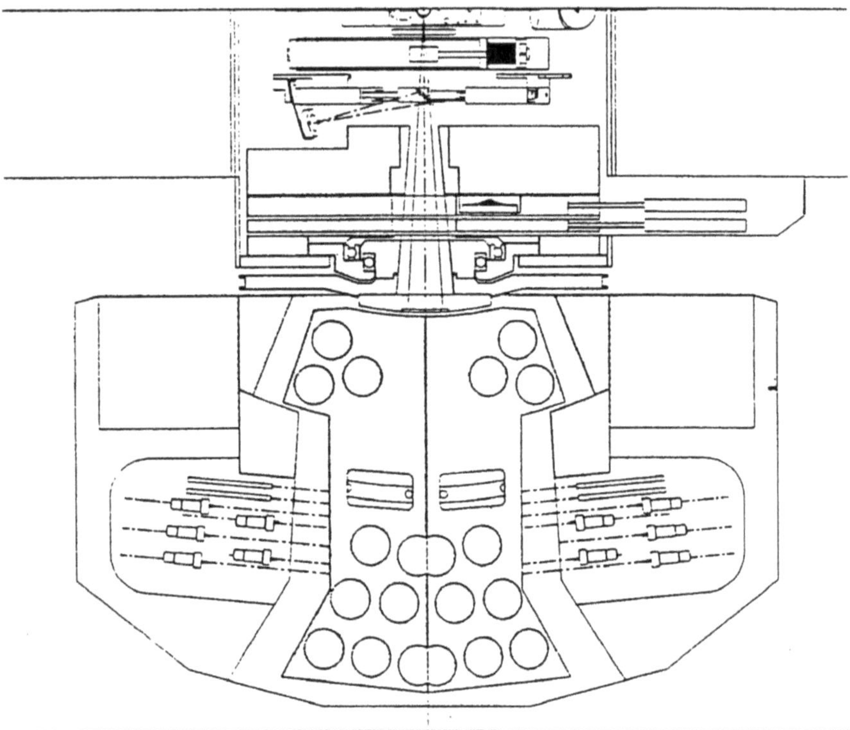

Fig. 6. View of the multileaf collimator at Louvain-la-Neuve. (From Denis et al. 1996)

be varied with the changing of inserts. Such a system has been used at Louvain for more than 10 years until the installation of a multileaf collimator in 1991 (Vynckier et al. 1993). Table 2 shows the collimation systems used at present in the neutron therapy centres.

Two different systems are worth discussing in detail: the multileaf collimator as used in Seattle (designed by Scanditronix, Brahme et al. 1983) and also adapted for the situation in Louvain-la-Neuve and the multirod collimator used in Detroit. Figure 6 illustrates a cross view of the multileaf collimator in Louvain-la-Neuve as described by Denis et al. (1996). It consists of a fixed pre-collimator and a moveable multileaf part. The pre-collimator is made of iron and defines the maximum field size of 22 cm×22 cm. The second part contains 44 individual motorized and computer-controlled leaves with a thickness of 1 cm and a length of 95 cm. These leaves are made of iron, containing 9 discs of borated polyethylene. For the clinical p(65)+Be neutron beam a transmission was measured of 10^{-3} neutrons when the collimator was completely closed. The other dosimetric characteristics are illustrated below.

Figure 7 illustrates a view of the multirod collimator as described by Maughan et al. (1996). This multirod collimator consists of two parallel op-

Fig. 7. The multirod collimator at Detroit. (From Maughan et al. 1996)

posed arrays of tungsten rods. The rods are supported by steel plates and can be pushed into the neutron beam by two high-density polystyrene foamblocks. These foamblocks can be cut out into the irregular field needed for therapy. Eighty-four arrays are arranged in 4 arrays of 21 rods each, with an equivalent thickness of 6.5 cm. Maximum field sizes up to 26.5×30 cm can be achieved at the isocentre. Each row contains 64 rods, bringing the total number of rods to 10840. Completely closed the neutron beam is collimated by 84 layers of rods, bringing the attenuation to 3.5×10^{-2}. The resulting dosimetric characteristics are also illustrated below.

Dosimetric Characteristics of Clinical Neutron Beams

Clinical neutron beams should present comparable dosimetric characteristics to those obtained for photon beams with linear accelerators. The penetration of the neutron beam should at least be comparable with the penetration of a 4-MV photon beam or higher. Due to scattering, neutron beams will present a higher scatter component as a function of depth (even higher than the scatter component of a cobalt-60 beam), resulting in a higher scatter under shielded areas and a larger penumbra. In order to be competitive with conventional photon beam therapy, the characteristics of the neutron beam should not deviate too much from the photon beam characteristics. Moreover, the dose rate should [corrected with the clinical relative biological effectiveness (RBE)] also be comparable with the typical dose rate of 2 Gy/min obtained with Linac photon beams.

Dose Rates Obtained with Neutron Therapy Machines

As already described, DT machines use a short SSD in order to obtain a reasonable dose rate. Indeed the dose rate is a limiting factor for these machines due to the heat production in the tritium target. Typically a dose rate of 0.15–0.20 Gy/min can be obtained at a distance of 1.25 m. Dose rates obtained for cyclotron-produced neutron beams were reviewed in ICRU 45 (ICRU report 45, 1989). The dose rate per unit of charge produced at a distance of 1.25 m can be presented as an exponential function as a function of the energy, yielding the following formulas: (a) for the deuterium on beryllium reaction:

$$D_{\mathrm{T}}/Q(Gy/C) = 2.49 \times 10^{-2} E_{\mathrm{d}}^{2.95}(\mathrm{MeV})$$

whereas (b) for the proton on beryllium reaction a comparable relation is obtained:

$$D_{\mathrm{T}}/Q(Gy/C) = 2.44 \times 10^{-2} E_{\mathrm{p}}^{2.49}(\mathrm{MeV})$$

For the deuterium on beryllium reaction a much higher dose rate will be obtained since more charge will be produced. Typically for the same current on target, the ratio of the dose rate of a d(50)+Be to a p(65)+Be neutron beam will be approximately 3.5. The resulting variations are also illustrated in Fig. 8.

Penetration

As for the dose rate, the penetration expressed by the percentage depth dose curves should be comparable with the penetration obtained by X-rays from modern linear accelerators. Also the penetration from existing neutron generators was reviewed by the ICRU (ICRU report 45, 1989) and expressed by the Z_{50}, the depth of the 50% isodose. This depth Z_{50} corrected to an SSD of 125 cm, could also be obtained by a simple expression:

$$Z_{50}(\mathrm{cm}) = 2.9\, E_{\mathrm{d}}^{0.40}\ (\mathrm{MeV})$$

for deuteron on beryllium generated neutron beams and:

$$Z_{50}(\mathrm{cm}) = 2.07\, E_{\mathrm{p}}^{0.51}\ (\mathrm{MeV})$$

or

$$Z_{50}(\mathrm{cm}) = 1.63\, E_{\mathrm{p}}^{0.53}\ (\mathrm{MeV})$$

for proton on beryllium generated neutron beams with or without a polyethylene hardening filter, respectively.

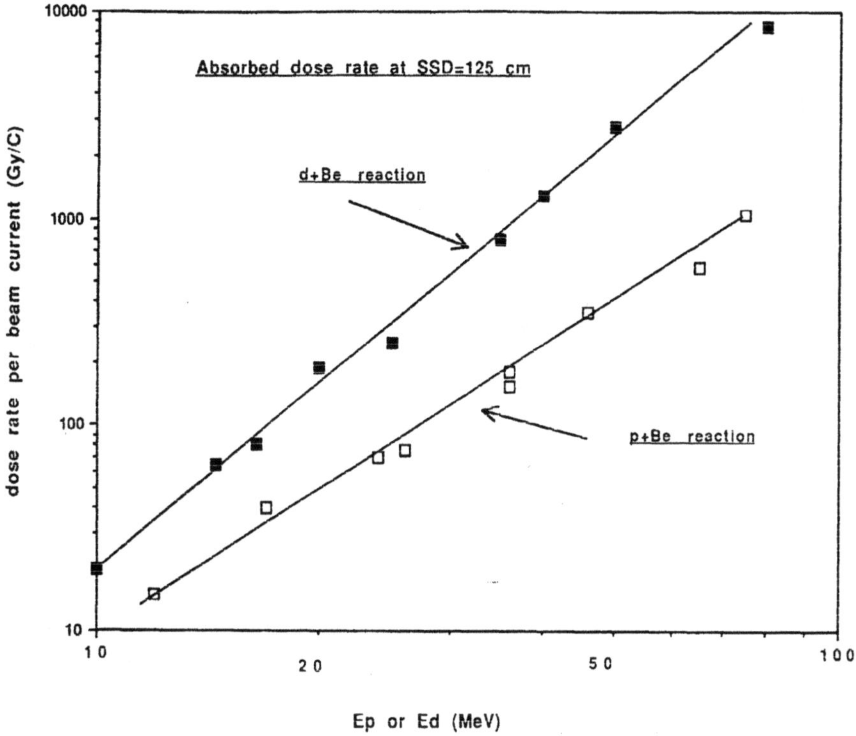

Fig. 8. Dose rate as a function of the energy for the deuterium and proton on beryllium reaction

Some typical percentage depth dose curves and the corresponding Z_{50} are also shown in Fig. 9 and Table 3. In first instance, the penetration of a p(65)+Be neutron beam compares closely to the penetration of an 8-MV photon beam, whereas the penetration of the p(35)+Be neutron beam is close to the penetration of the cobalt-60 photon beam. The penetration of the d(50)+Be neutron beam will be approximately between these values. However, these data are difficult to compare since the penetration is SSD dependent and the variation of the penetration of a neutron beam will vary more with field size than a photon beam due to the larger scattering component.

Measurements of the percentage depth dose curves should be performed in a water phantom. Water does not present the correct density or composition of tissue. However, the attenuation and scattering properties of neutron beams in water correspond closely. On the other hand, a scaling method is available if other phantom materials are used (for instance for daily practice) (Vynckier et al. 1985). Built-up curves of a neutron beam are comparable to the built-up curves of photon beams. When compared to the built-up curves of photon beams, they will, however, show a slightly higher dose at zero depth, a faster increase to the maximum but a flatter maximum. The maximum depth will be 2 cm and 1.5 cm for a p(65)+Be and a d(50)+Be neutron beam, respectively.

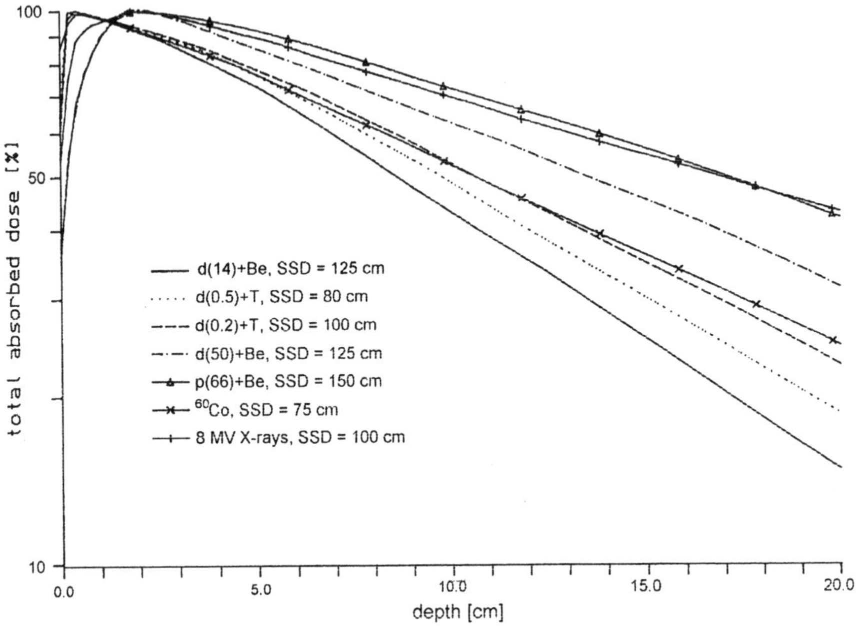

Fig. 9. Depth dose distributions from DT generators, low-energy cyclotrons d (14)+Be, and high-energy cyclotrons [p (66)+Be], compared with photon therapy facilities (Rassow et al. 1992)

Table 3. Comparison of dosimetric data of some clinical neutron beams

Centre	Reaction	SSD (cm)	Z50 (cm) 10×10 cm	Penumbra (cm) $z = 10$ cm	I_{SURF}/I_{MAX} (%)
Seattle	p(50)+Be	150	14.8	1.40	38
Clatterbridge	p(62)+Be	150	16.2	1.60	33
Fermilab	p(66)+Be	190	16.6	2.00	40
Faure	p(66)+Be	150	13.6	1.25	33
Detroit	d(50)+Be	182.9	16.5	1.70	44
Louvain	p(65)+Be	182.5	17.6	1.25	61
Hamburg	d(0.5+T	80	9.0	2.4	55
	Co-60	80	11.6	1.60	40
	RX 8MV	100	17.1	0.80	33

Penumbra

Also, due to the larger scattering component, the penumbra of the neutron beams will generally be larger than the penumbra of photon beams. Table 3 compares the penumbra at 10 cm depth of clinical neutron beams. They are all between 1.5 and 2 cm for cyclotrons whereas the penumbra of X-rays from a linear accelerator is less than 1 cm at that depth. With the new multileaf collimator at Louvain-la-Neuve, a penumbra of 1.25 can be achieved due to the long leaves of 95 cm. However, as a function of depth these penumbra will increase much faster than those of the corresponding X-rays.

Neutron Beam Measurements

Neutron Spectrometry

For dosimetric purposes the energy spectrum of the neutron therapy machine in the phantom needs to be known. Very common in neutron spectrometry is the time-of-flight (TOF) technique, where the time between generation and detection in a counter is determined and converted into the kinetic energy. This requires a pulsed neutron source and sufficient space between source and detector. In-phantom measurements are not possible. TOF measurements are used in standard neutron dosimetry laboratories for the determination of cross sections and dosimetry.

For in-phantom measurements two different techniques are used: activation and proton-recoil spectrometry. For activation spectrometry the detectors must have different well-known activation cross sections as a function of neutron energy. The activation detectors are exposed at identical positions in a phantom. The neutron-induced activation of the applied reaction must be determined.

In the thermal neutron energy region, activation detectors are used with and without filtering by Cd. At the energy of large resonances in the epithermal energy region self-absorption effects are employed by detectors that are used in sandwich assemblies.

In the fast neutron energy region, detectors with different thresholds of the activation cross sections are used to determine the fast neutron energy spectrum. From the measured activation and the cross section data a fast neutron spectrum is iteratively determined. Usually a first estimate of the spectrum has to be given. From this "guess" spectrum the activation is simulated and the difference to the measured values is calculated. The spectrum is then iteratively modified until the differences are below an acceptable value. Proton recoil spectrometers can be used in phantoms, too. The neutron-sensitive detector is usually an organic liquid (NE 213) or crystal (Stilbene) detector, mounted via a light-guide onto a photomultiplier. Scintillation pulses are generated by neutrons and photons. The different shapes of the pulses available at the photomultiplier make a neutron-photon discrimination possible. The accumulated pulse-height spectra are analysed by unfolding procedures. A common method is to generate a response matrix of the detector by Monte-Carlo simulations of mono-energetic neutron sources taking into account all possible reactions, multiple scattering, wall effects, light-yield functions, etc. The appropriate size of a scintillation detector is dependent on the maximal range of the protons. For DT neutrons small detectors 0.5″ in diameter and 0.5″ in length were used with satisfactory results. In

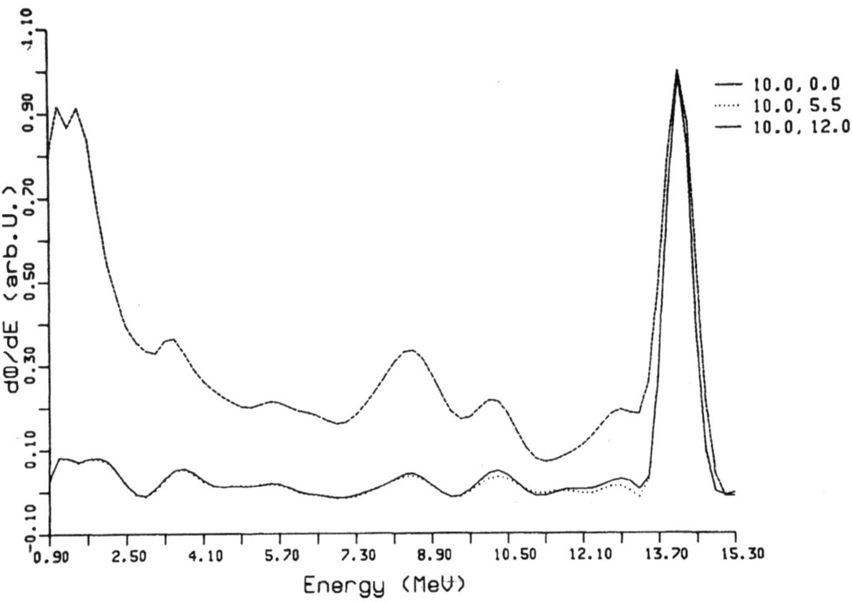

Fig. 10. Neutron spectra of a DT generator at 10 cm depth in a water phantom: at the centre, at the penumbra, and outside a 10 cm×10 cm field

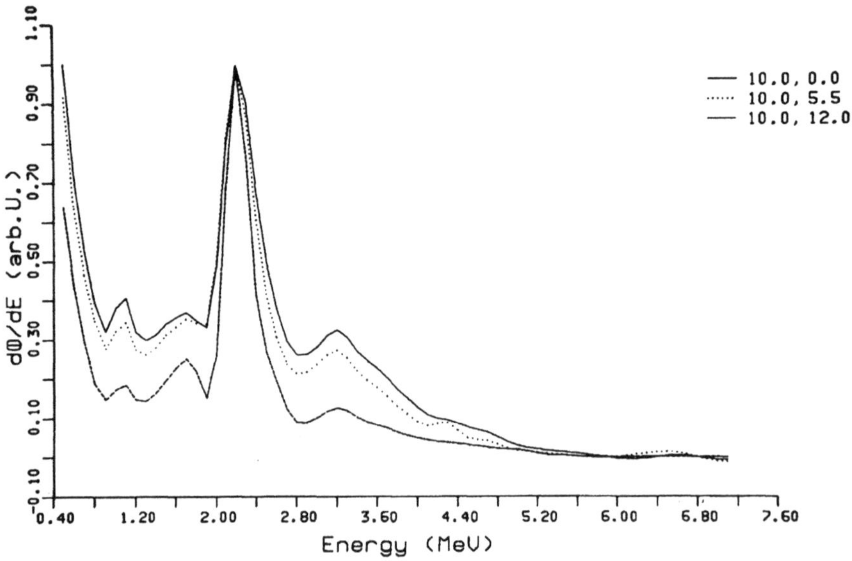

Fig. 11. Photon spectra at the same positions as in Fig. 10

Fig. 10 the unfolded neutron spectrum of a DT generator in different positions in a water-phantom inside and outside the beam at 10 cm depth is shown. The spectra are normalized to a unique peak-maximum.

As the scintillator responds to photons as well, the discriminated pulses can be used to evaluate the corresponding photon spectra. In Fig. 11, a relative high intensity of photons with 2.2 MeV energy arising from the capture of thermal neutrons by hydrogen is seen. This dominant capture process is responsible for the fact that the photon energy spectra are less dependent on the position than the neutron spectra and are even comparable for different neutron sources (Schmidt and Hess 1982).

Microdosimetry

The biological effect of radiation is due in part to the local energy deposition of charged particles in microscopic sites of tissue. When neutrons react with tissue they produce a broad spectrum of charged particles with a different LET. The LET spectrum of neutrons and photons can in principle be calculated. An approximation to the LET spectrum can be otained by experimental investigations with tissue-equivalent proportional counters. The measured microdosimetric spectra in terms of the lineal energy y are obtained by multiplying the dose frequency $d(y)$ with y. There are some restrictions involved with the microdosimetric spectra (ICRU 1982). For practical reasons the volume of the energy deposition is spherical. The density of the counter gas can be chosen so that the energy loss of the charged particles traversing the counter is equivalent to the energy loss in 1 μm tissue. If the gas cavity of the counter is small in comparison with the range of the charged particles, and if effects of straggling can be neglected, the microdosimetric spectra give a good representation of the LET distribution. The microdosimetric spectra are a superimposition of the different types of charged particles. In Fig. 12 the microdosimetric spectra of 14 MeV neutrons at different depths on the beam axis and at a position outside the useful beam in a solid TE phantom are given. The main peak arises from protons, the lower part of the spectrum is due to electrons, and the upper part is due to α-particles and heavy recoils. By use of proportional counters with different wall material, the components of even size spectra can be evaluated (Schmidt and Hess 1988). The separated contribution of the charged particles to the neutron dose can be calculated from microdosimetric spectra. Based on several measurements at different lateral positions in a constant depth in a solid TE phantom, the neutron dose was separated into the contributions from protons, α-particles, and heavy recoils in a 14-MeV neutron beam (see Fig. 13).

Extensive work on the microdosimetric intercomparison of the European neutron therapy centres was performed by Pihet et al. (1990). Microdosimetric spectra and radiobiological data were obtained under identical conditions. These data were used to evaluate a biological weighting function (RBE) versus lineal energy in order to provide a solution to account for radiation quality

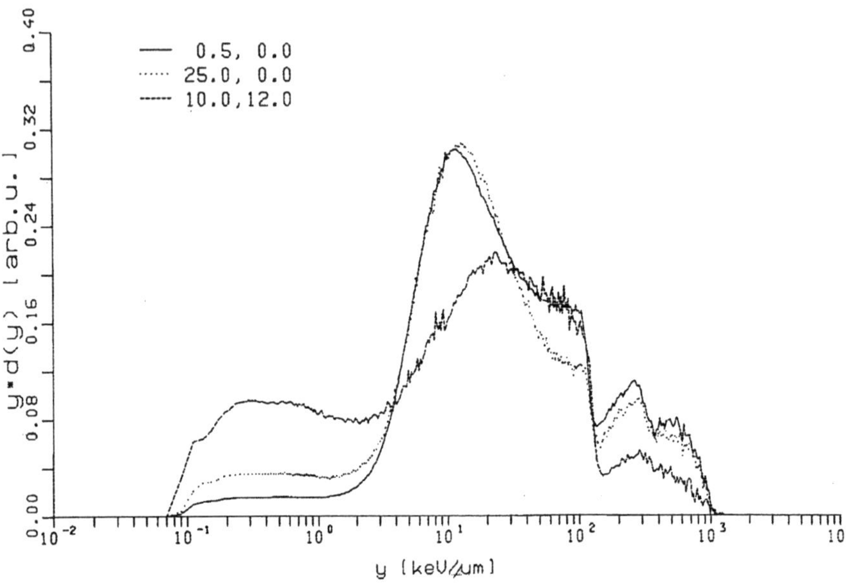

Fig. 12. Microdosimetric spectra of a DT generator at different positions in a TE phantom

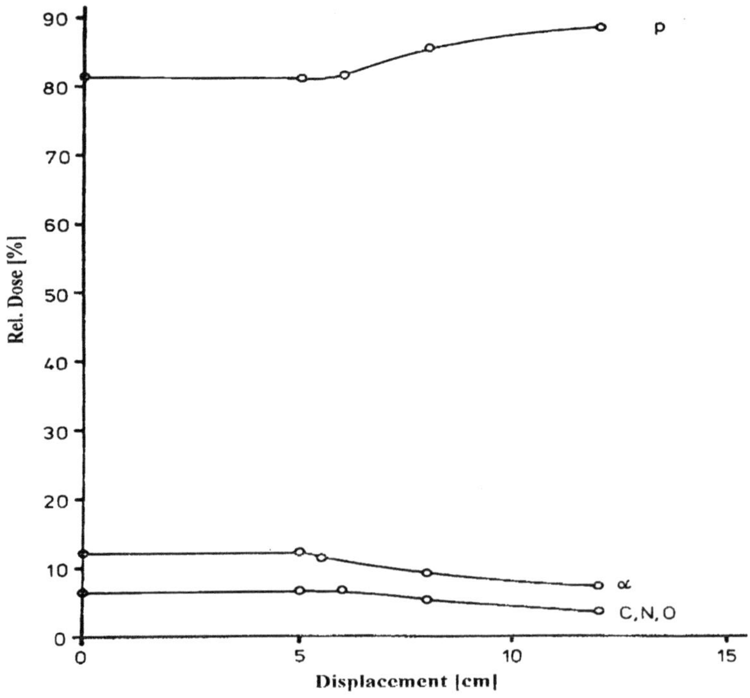

Fig. 13. Separation of neutron dose into the contributions from protons, α-particles and heavy recoils at different lateral positions in a constant depth in a solid TE phantom. Field size, 10 cm×10 cm, 14 MeV neutrons

variations. An impression of the wide range of microdosimetric spectra used in radiobiology and radiotherapy with neutrons can be otained from Fig. 14. The biological weighting function obtained with investigations for tolerance of mouse intestinal crypts is given in Fig. 15. From these data biologically equivalent doses can be derived. In Table 4 the participating institutions with the neutron-producing reaction are given. The values for the dose per fraction leading to comparable biological effects are listed, either in comparison with 1.0 Gy/fraction with p(65)+Be neutrons, or with 0.8 Gy/fraction of dT neutrons.

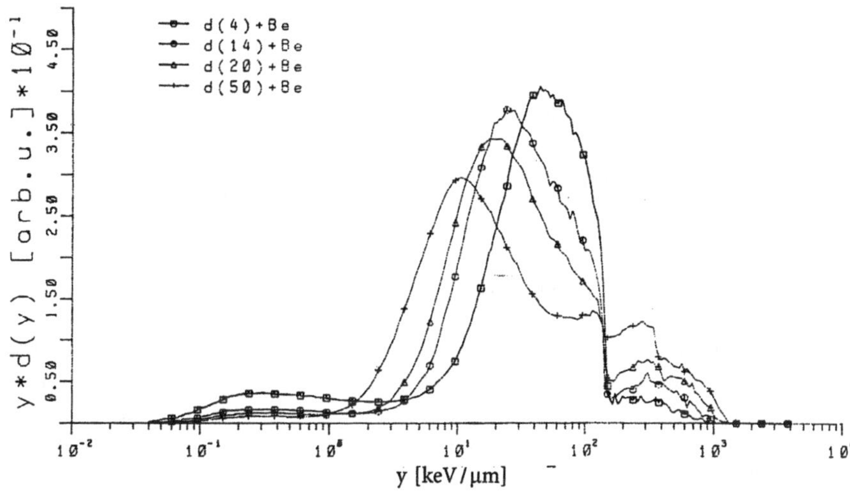

Fig. 14. Microdosimetric spectra at different European neutron beams, measured by Pihet et al. (1990)

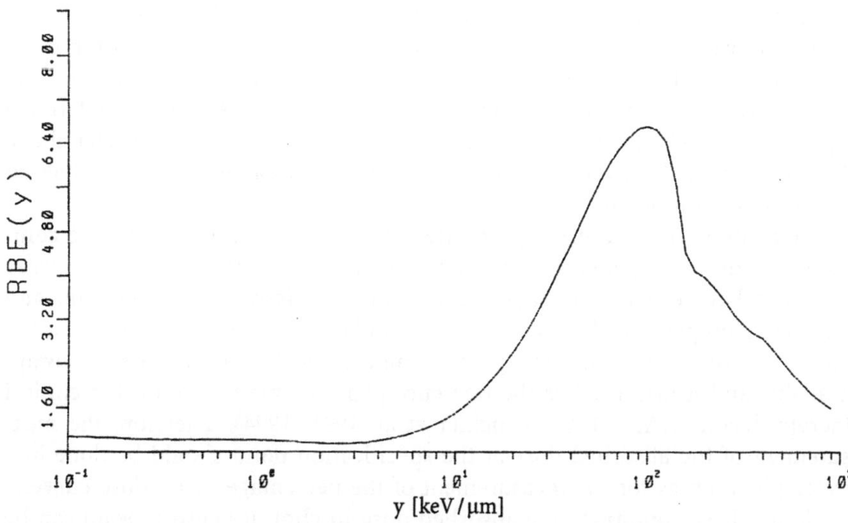

Fig. 15. Biological weighting function evaluated from microdosimetric spectra and RBE data

Table 4. Biological equivalent doese per fraction resulting from the radiobiological intercomparison

Reaction	Dose per fraction compared to		
	Institute	1.0 Gy for p(65)+Be neutrons	0.8 Gy for DT neutrons
d(4)+Be	Gray Lab.	0.49	0.51
d(14)+Be	Essen	0.58	0.60
d(14)+Be	Gent	0.61	0.63
p(34)+Be	Orleans	0.89	0.92
p(34)+Be	Louvain	0.81	0.84
p(45)+Be	Louvain	0.86	0.89
p(50)+Be	Louvain	0.93	0.97
p(62)+Be	Clatterbridge	0.97	1.01
p(65)+Be	Louvain	1.0	1.04
d(.5)+T	Hamburg	0.77	0.80

Determination of the Total Absorbed Dose

Dosimetry for clinical neutron beams was the subject of discussion in the 1980s (Broerse et al. 1981; Mijnheer et al. 1987). However, the dosimetry of the clinical neutron beams has been well established since the publication of the ICRU Report No. 45 (ICRU 1989). This chapter will describe briefly the techniques and methods recommended for the determination of the total absorbed dose both under reference conditions and for relative measurements. The measurement of the gamma component will be added.

The reference dosimeter for the determination of the total absorbed dose in neutron beams is a tissue-equivalent ionization chamber made of A-150 tissue equivalent plastic. Two models fall within the criteria for neutron dosimetry (i.e. sensitivity and size) and are accepted as reference dosimeter: the FWT IC17 tissue equivalent ionization chamber and the Exradin T2 tissue equivalent ion chamber. Physical data will only be given for both chambers. A formalism will only be given for ion chambers calibrated in a cobalt-60 beam in air in terms of air-Kerma.

Attenuation and scattering properties of clinical neutron beams were compared for different phantom materials (Vynckier et al. 1985). Although water does not have exactly the same composition as tissue, the attenuation and scattering properties of water are very similar to tissue with compensation due to the differences in density and composition. Moreover, water is always available and is also used as the reference phantom material for other clinical therapy beams (IAEA 1987; Vynckier et al. 1991, 1994). Therefore the measurement of the absorbed dose at the specification point should be done in a water phantom as for the measurement of the percentage depth dose curves.

Under these conditions the absorbed dose in clinical neutron beam can be obtained by the following expression:

$$D_T = M_T \cdot (\Pi k_M)_T \cdot N_K \cdot A_{wall} \cdot (1-g) \cdot [(\mu_{en}/\rho)_t/(\mu_{en}/\rho)_{air}]_c \cdot d_T \cdot (1-\delta)^{-1} k_T^{-1} .$$

The different terms in this formula have the following meanings: M_T is the electrometer reading; $(\Pi k_M)_T$ is the product of several correction factors in order to correct the elctrometer reading to the real charge produced; N_K is the air-Kerma calibration factor; A_{wall} is the correction of the chamber response for absorption and scattering during calibration in air; g is the fraction of energy of secondary particles converted to bremsstrahlung in air; $(\mu_{en}/\rho)_t$ is the mass energy absorption coefficients for material t; d_T is the displacement correction factor; $(1-\delta)^{-1}$ is the correction for the difference in response of the TE ion chamber for neutrons compared to photons; and the expression of k_T is given by:

$$\frac{1}{k_T} = \left[\frac{(r_{m,g})_n}{(\bar{L}\rho)_m/(\bar{L}/\rho)_g}\right] \cdot \frac{W_n}{W_c} \cdot \left[\frac{(K_t/K_m)_n}{(\mu_{en}/\rho)_t/\mu_{en}/\rho)_m}\right]_c$$

where $(r_{m,g})_n$ is the gas-to-wall absorbed dose conversion factor; $(\bar{L}/\rho)_m$ is the mean restricted mass stopping power in metres; W_n and W_c are the average energy expended per ion pair formed in the neutron and photon beams, respectively; and K_t is the neutron Kerma in material t (tissue).

Values for these physical parameters have been extensively discussed in the past. The values recommended in ICRU 45 (ICRU 1989) are given in Table 5. With this method the overall uncertainty (1 SD) for the determination of the absorbed dose in a high-energy neutron beam will be 4.6%.

Table 5. Values of the physical parameters recommended for neutron dosimetry with tissue equivalent ion chambers

Source	Symbol	Value
Wall correciton factor	A_{wall}	0.983
Ratio of the average energy required to form an ion pair for neutrons to the calibration beam	$\frac{(W_g/e)_n}{(W_g/e)_c}$	1.06
Ratio of the photon mass absorption coefficients	$[(\mu_{en}/\rho)^t_{air}]_c$	1.02
Ratio of the mean restricted collision mass stopping powers	$[(\bar{L}/\rho)^m_g]_c$	1.00
Gas-to-wall absorbed dose conversion factor	$(r_{m,g})_n$	1.00
Fraction of energy converted to bremsstrahlung	g	0.003
Ratio photon to neutron response	δ	<0.01
Neutron Kerma ratio	$(K_t/K_m)_n$	0.950
Displacement factor (Exr. T2)	d_T	0.986
Displacement factor (FWT IC17)	d_T	0.978

Determination of the Gamma Component

The gamma component is generally low and of the order of a few percent in the high-energy neutron beams [p(45)+Be or higher] (Vynckier et al. 1987). This component can be determined using two methods. The first is based on the use of two independent detectors: one which is sensitive to the neutrons (generally the tissue equivalent ionization chamber) and one which is insensitive (or has a low sensitivity) to neutrons. The insensitive detector is generally a non-hydrogenous ionization chamber (C-CO_2 or Mg-Ar) or a GM counter. The latter is preferred since its neutron sensitivity (k_U) is smaller and less variable (several percent) than the sensitivity of the non-hydrogenous chambers. k_U values for GM counters in different neutron beams were reviewed by Pihet et al. (1982).

A second method is based on microdosimetry as explained above. The analysis of the microdosimetric spectra and the comparison of the low LET part (<10 keV/μm) allows the determination of the gamma component. Both methods were also reviewed by Pihet et al. (1984). Figure 16 illustrates a typical variation of the gamma component as a function of depth in different neutron beams.

Treatment Planning for Neutron Therapy

For the application of neutrons in radiotherapy the distribution of absorbed dose in the patient has to be determined. As different components in a neutron field have different RBEs, the total absorbed dose needs to be separated

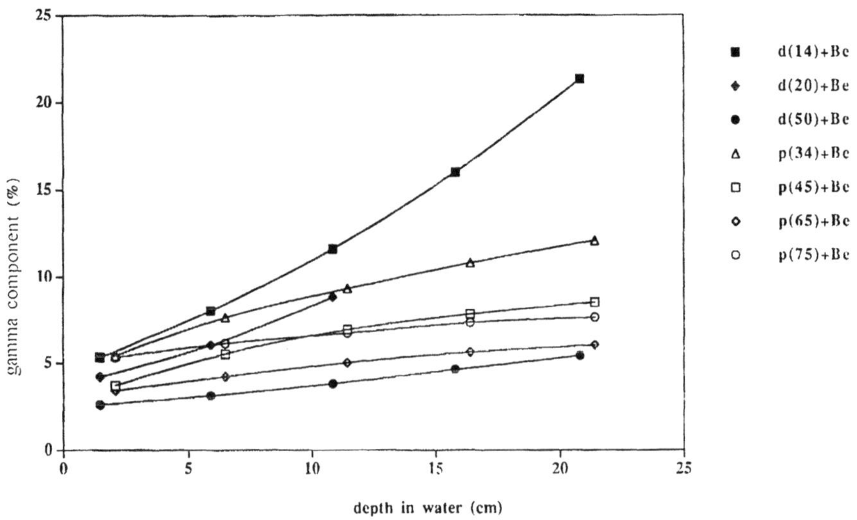

Fig. 16. Variation of the gamma component (D_G/D_T) as a function of the depth in a phantom for different clinical neutron beams

into the neutron and photon components, and, if essential for the dose, into the low-energy neutron doses, too. Different mathematical models can be used for the dose calculation in the patient. But before this can be done a set of dose distributions has to be measured under reference conditions either for direct input into the treatment planning algorithm or for the determination of parameters for generating functions.

Generation of Absorbed Dose Distributions for Neutron Beams

The distributions of the total absorbed dose are commonly determined by use of TE/TE ionization chambers, assuming equal response of the chamber to neutrons and photons. For the determination of the relative photon component a second detector needs to be introduced. This detector must be less sensitive to neutrons, i.e. an Mg/Ar ionization chamber or a GM counter. The use of GM counters is complicated by the fact that dead-time corrections have to be made and consequently a low dose rate is required, whereas the use of an Mg/Ar chamber can be performed at the same dose rate and with the same equipment as applied for the primary TE/TE chamber. For the determination of dose distributions the use of a water phantom is recommended (Broerse et al. 1981). The same phantoms applied in photon dosimetry can be used, replacing only the ionization chamber.

Central Axis Depth Dose Distributions

Relative absorbed dose distributions are usually expressed in percent, relative to the maximum or to a reference depth. An essential part of the depth dose distribution is the dose build-up. The shape of the build-up is dependent on the neutron spectrum, the target thickness, collimating system, and accessories such as filters and wedges (ICRU report 45, 1989). Generally the build-up has to be measured separately, preferably by use of an extrapolation chamber. The depth-dose curves are mainly dependent on the neutron-producing reaction and the SSD. Some typical depth dose curves of therapeutic neutron machines are given in Fig. 9. For comparison the data for ^{60}Co and for 8-MV X-rays are included, indicating that high-energy cyclotrons generate a depth dose characteristic comparable to Linac-generated photons.

Transverse Dose Distribution

Transverse dose distributions are either related to the dose at the maximum on the central axis or to the dose on the central axis at the measured depth. From transverse dose distributions detailed information about the flatness of the field, the penumbra and the dose outside the beam are obtained. For treatment of patients it is generally desirable to have total flatness, a steep dose gradient in the penumbra and a negligible dose outside the beam. A major parameter influencing dose distributions is the neutron spectrum and the directional dependence of the neutron intensity. Neutrons generated from the DT reaction are isotropically emitted, so that their characteristics can be compared with an intense γ-source. Cyclotron-produced neutrons are preferably generated in a forward direction, with an anisotropy increasing with energy. This can be compared to X-ray production with linear accelerators. For high-energy neutrons a flattening filter is required to compensate for the anisotropy.

The penumbra is dependent on the size of the focus, the collimator properties and, again on the neutron energy. Generally the penumbra width decreases with increasing neutron energy (Mijnheer and Broerse 1979). Outside the beam the relation between neutron and photon dose changes; the neutrons are usually more collimated than the photons. This leads to a decrease in the RBE outside the beam compared to the RBE of the primary beam.

Isodose Distribution

Isodoses for reference conditions can either be calculated from central axis depth dose and transverse dose distribution or can be measured directly in terms of total absorbed dose. The direct measurement, either via "isodose hunting" or by two-dimensional net scans, can be used to control the generation of isodoses from depth dose and beam profiles. In Fig. 17 the corresponding comparison for a 10 cm×10 cm field with DT neutrons is performed.

Different mathematical models have been developed for the generation of isodose distributions, often adapted from photon calculations (Rassow et al. 1984). The straightforward method uses a two- or three-dimensional array of directly measured dose values. The accuracy is dependent on the grid size of the matrix. The matrix may have a constant grid or a variable grid with a

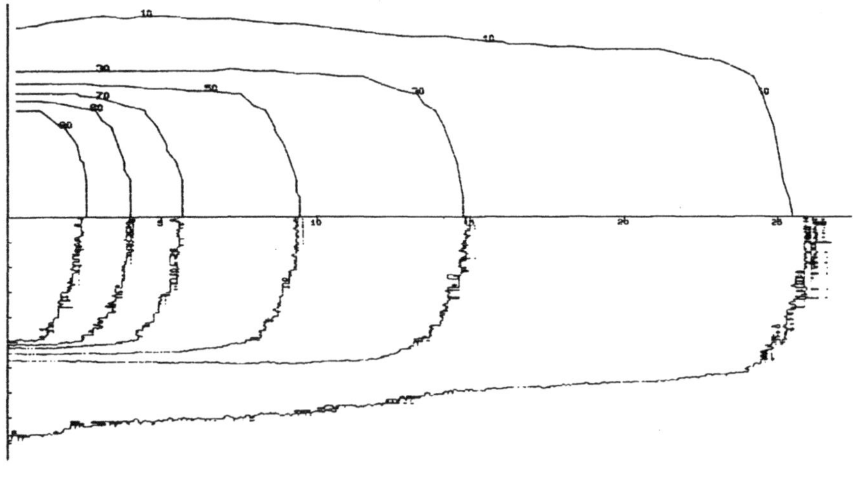

Fig. 17. Comparison of calculated (*upper*) with measured (*lower*) isodoses

gradient-dependent spacing, possibly using fan lines according to the divergence of the beam. Besides directly measured data, a matrix can be generated from depth dose and profile scans. For some treatment units, the matrix can be reconstructed by use of the decrement line method. If the decrement lines are not straight, the accuracy is determined by the number of profile scans and the curvature of the decrement lines. Another approach uses generating functions, either purely empirical or based on physical and geometrical properties. The set of parameters has to be obtained from dose measurements and dose distributions for different field-sizes and SSDs.

According to a well-konwn TAR/TPR method for photons, equivalent models for neutrons are available. The absorbed dose is separated into a primary component, stated as TAR/TPR for field-size zero, and the scatter component, stated as scatter-air- or scatter-phantom-ratio. For photons TAR/TPR values are generally independent of SSD whereas for neutrons an additional correction may be necessary to obtain agreement at different SSDs.

Another approach uses theoretical models, based on the solution of the transport equation for neutrons. Though the calculations require a very powerful computer, the solutions are satisfactory. Even Monte Carlo calculations are used for the determination of dose distributions. The direct use of these methods in treatment planning is still limited by computer time. On the other hand, these procedures can be used to generate data for the models described above, for instance, the generation of primary and scatter dose components.

Absorbed Dose in a Patient

Treatment planning with fast neutrons means the calculation of a two- or even three-dimensional dosematrix in a heterogeneous medium. In contrast to reference conditions during treatment, the beam is directed toward an irregularly shaped surface at angles which are often not rectangular. The corrections applied for irregularly shaped surfaces in treatment planning with photons can generally also be used with neutrons. The isodose shift method originating from manual isodose generation consists of shifting the isodose according to the missing or additional tissue compared to an orthogonal surface. The applied shift factors depend on the neutron spectrum and are generally less than 1. In Fig. 18, a comparison of measured and reconstructed isodoses for an oblique field is given (Vynckier et al. 1981). The agreement of the results is comparable to photon treatment planning. The effective SSD method is more common with computer-based treatment planning. The results, given as an example in Fig. 18, are also satisfactory. If the primary and scatter components are analysed, the TAR method may be used. As mentioned above, corrections for the SSD dependence might have to be introduced for a neutron beam.

The dose generation procedures described above are used to calculate the dose in the patient. The heterogeneity needs to be taken into account. For

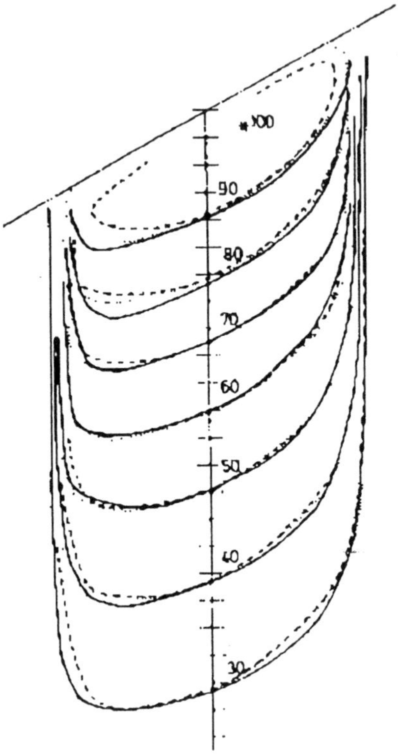

——— measured in water
-------- effective SSD method
.... isodose shift method

Fig. 18. Comparison of calculated and measured isodose distributions (from Vynckier et al. 1981)

photons the conversion of depth dose data from one medium to another is often performed proportional to the density. For low- and high-energy neutron beams it has been shown that scaling factors can be used to convert the depth dose from one medium to another (Awschalom et al. 1983; Hensley et al. 1985). For lung and air cavities the effect can be taken into account by use of the effective attenuation method.

Besides this transmission calculation, a Kerma correction needs to be applied to take into account the difference in atomic composition of soft tissue and the heterogeneity. The proton contents of the organs are essential for the energy transferred to the tissue. As the atomic composition in a transverse slice of a patient is not known, additional information may be given by classifying the visible organs so that the Kerma correction can be taken from a table or by directly converting Hounsfield data into Kerma corrections (Schmidt and Thom 1990). Both procedures are rough approaches. On the other hand, the Kerma corrections are within 0.7 (compact bone) to 1.14 (fat) and are slightly dependent on the neutron spectrum.

Dose Specification

For radiotherapy with neutrons the recommendation of the ICRU for prescribing, recording and reporting photon beam therapy can be adapted with minor modifications (ICRU 1994). The main differences arise from the fact that neutrons are always contaminated by photons with a different RBE. Therefore the reported doses at the ICRU reference point should include the total dose (neutron+photon) and the photon dose. Outside the neutron beam the relation between neutron and photon dose changes so that a decrease in RBE occurs. Therefore in a three-dimensional treatment plan with neutrons the photon dose at the organs at risk should also be given. As the absorbed dose in a patient depends on the proton contents, the proton contents at the ICRU reference point should be representative of the planning target volume (PTV).

The quality of the neutron dose distribution discussed above may result in a less homogeneous dose in the PTV than with the modern linear accelerator-generated photons. Nevertheless the difference between the maximum and the minimum dose in the PTV should be within 20%.

In treatment regimes where both neutron and photon treatments are applied, the dose distributions for each PTV need to be calculated, the neutron and the photon doses need to be added separately, and the doses at the reference points need to be stated separately as the doses from neutron and photon treatment.

Three-Dimensional Treatment Planning

The common basis for neutron treatment plans is computer topography (CT) slices. For three-dimensional treatment planning, coplanar CT slices must be used. Each of these slices needs to be analysed to draw the PTV and the organs at risk. Additional information can be obtained by simultaneous use of magnetic resonance imaging (MRI). If an appropriate MR sequence is used, the proton contents may be obtained from MRI data. Many tools adapted from three-dimensional treatment with photons are used, i.e. beam's-eye view (BEV) and dose volume histograms (DVH).

If the algorithms used in the planning system enable a separation of the dose into the primary and the scatter components, irregular-shaped fields can be used. The shape of the field can be determined by use of external absorbers in regular fields, or by application of multileaf or multirod collimators. The lack of scattered radiation in the open area and the radiation scattered into the shielded region have to be calculated, either directly by use of pencil beam simulation or indirectly by modifying the open field dose distribution. Algorithms well known from photons can be adapted with more or less complicated modifications, i.e. the sector integration method (Clarkson 1941), the area to perimeter correction (Wrede 1972) or the ring model (Jacobs annd Van Kleffens 1986).

The use of irregularly shaped fields from different non-coplanar ports is often called conformation therapy, because the three-dimensional shape of the dose distribution conforms to the shape of the target. The typical three-dimensional tools (BEV, DVH) are independent of the type of radiation. The newly developed neutron therapy machines are prepared for conformation therapy.

Isocentric neutron therapy units are capable of performing dynamic therapy. The depth penetration of DT neutrons requires multiple isocentric fields or rotational therapy for deep-seated tumours. The shape off the isodoses can be modified by changing either the rotational speed or the dose rate. By use of this weighted beam rotational therapy (WBRT), a variety of dose distributions conforming to the target is possible, and the parameters can be obtained from a computer-controlled optimization procedure (Höver et al. 1990).

By use of small neutron fields a convergent beam irradiation well known from stereotactic photon therapy may be performed. Several non-coplanar isocentric rotational fields are applied to generate a spherical homogeneous dose concentration with a steep gradient.

As Monte Carlo codes for photons and for neutrons are still too time consuming for the generation of three-dimensional dose distributions in a patient, parameterized functions are commonly used for three-dimensional treatment planning. The basis of these functions is formed by point spread functions or kernels generated by Monte Carlo codes, where the parameters are adapted to the individual particle spectrum. It is possible to generate point spread functions and kernels for both neutrons and photons as suitable codes are available. The heterogeneity correction for neutrons is more complicated than for photons. But if the density-scaling or collapsed-cone algorithms used for photons are modified to take into account the attenuation of neutrons, these algorithms can also be adapted. In addition the Kerma correction, mainly based on the proton contents of different tissues; has to be applied. This can be supported by additional use of MRI data that represent the proton density.

Whenever the dose matrix is calculated for the fast neutrons, the same display modes as used in photon treatment planning can be used. For three-dimensional treatment plans the visualization includes the topography, volumes of interest and dose distribution. Wire-frame graphics is often used to make a fast display possible. But more sophisticated visualization can use complex three-dimensional rendering techniques with shaded surfaces and partial transparency. Anatomical structures such as bone and lung can automatically be segmented from CT data, whereas artificial structures such as the target volume need to be rendered from a separate data set containing the information of the volumes of interest, i.e. target and organs at risk. The dose of certain level compared to the reference point can then be included in these structures. If the displayed dose level is continuously changed from lower to higher values, the effect of the dose concentrating onto the target is shown. An interactive rotation of the displayed structures helps to create a three-dimensional impression (Schmidt et al. 1992).

Biological Treatment Planning

For the assessment of a three-dimensional treatment plan, not only the physical dose distribution but also the biological effect of the dose on the target and organs at risk has to be taken into account. From the dose volume histograms the tumour control probability (TCP) and the normal tissue complication probability (NTCP) can be estimated. These calculations require biological parameters that are still not completely known for photons or for neutrons. Nevertheless these calculations result in numbers that can be used to compare different treatment plans and can be of help for optimizing the dose distribution (Goitein et al. 1987).

A different approach defines the relative biological dose (RBD) as the dose in relation to the tolerance dose of the organs at risk. Normalization of the RBD to 100% indicates the areas where the dose is equal to the tolerance dose. If this is displayed in three dimension, the biological treatment plan can be rendered in the same mode as the physical dose, showing the volumes with doses greater than the tolerance dose. Again the data are still not complete for the neutron treatment planning, but a comparison of different plans is possible and the three-dimensional display indicates the critical regions of a three-dimensional treatment plan for neutrons (Schmidt et al. 1994).

Conclusions

This paper has described the physical characteristics of neutron beams. It is possible to construct a hospital-based neutron therapy centre (i.e. the isocentric installation at Detroit or Seattle) which has the same technical performances as linear accelerators. However, at present neutron therapy is only performed at a limited number of centres. The results of the clinical trials of neutron therapy can only be compared, if from a technical, physical and dosimetric point of view the neutrons are comparable with conventional photon beams from linear accelerators. Quality assurance projects, including quality controls on the dosimetric and technical performance of the neutron therapy installations, in vivo dosimetry, and precision of the algorithms applied in neutron therapy should be at least as good as for photon beams. Moreover, the radiobiological characteristics of the neutron beam should be determined under reference conditions in order to allow comparisons to be made.

However, in order to be on the same level of precision, more problems need to be solved:

- The generation of these neutron beams is more complicated and will besides the standard quality control procedures need some supplementary quality verifications.
- In vivo dosimetry is well established in photon therapy. Different detectors (i.e. TLD) (Loncol et al. 1996) are available for neutrons, which will allow the initiation of an in vivo dosimetry project for neutrons.

- Quality control procedures for treatment planning systems are available for photon beams. Neutron therapy planning is mostly performed by adoption of the existing algorithms for photon beams to neutron beams. However, in order to improve the precision, more companies are providing systems that are utilizing algorithms related to the basic interactions of the ionizing beam (i.e. Helax). These systems can only be adapted for neutron beams if the basic calculations (Monte Carlo calculations for point spread functions) are available. Moreover, the problem of the corrections for heterogeneities is very specific for neutrons and can only be done in a correct way by the implementation of the correct physical interactions. Therefore data should be available on the cross sections of neutrons of high energy in basic elements, but also data should be available on the proton concentration in different tissues of the patients to be treated.

Finally the influence of the radiobiological properties of neutrons on different tissues should also be determined by the implementation of biological planning in order to provide the clinician with a realistic picture of the results of neutron irradiation in patients.

References

Awschalom M, Rosenberg I, Ten Haken RK (1983) Scaling neutron absorbed dose distributions from one medium to another. Med Phys 10:436–443

Brahme A, Enmaa J, Lindbäck S, Montelius A, Wootton P (1983) Neutron beam characteristics from 50 MeV protons on beryllium using a continuously variable multi-leaf collimator. Radiother Oncol 1:65–76

Broerse JJ, Mijnheer BJ, Williams JR (1981) European protocol for neutron dosimetry for external beam therapy. Br J Radiol 54:882–898

Clarkson JR (1941) A note on depth doses in fields or irregular shape. Br J Radiol 14:265–268

Denis JM, Meulders JP, Lannoye E, Longree Y, Reyckewaert G, Richard F, Vynckier S, Wambersie A (1996) The multi-leaf collimator for fast neutron therapy at Louvain-la-Neuve. Bull Cancer Radiother 83 Suppl 1:160 s–169 s

Franke HD, Hess A, Schmidt R (1983) The DT neutron generator for tumour therapy in Hamburg-Eppendorf. Atomkernenergie Kerntechnik 43:95–104

Goitein M, Urie M, Munzenrider JE, Gentry R, Lyman JT, Chen GTY, Castro JR, Maor MH, Stafford PM, Sontag MR, Altschuler MD, Bloch P, Chu JCH, Richter MP (1987) Evaluation of treatment planning for particle beam radiotherapy. Report prepared for National Cancer Institute

Hensley FW, Rassow J, Temme A (1985) Tissue equivalence in clinical neutron dosimetry: comparison of dose distribution in nine tissue substitutes for a d(14)Be neutron beam. Med Phys 12:350–356

Höver KH, Hartmann G, Lorenz WJ, Menzel HG, Schraube H, Wolber G (1975) Dosimetry at the neutron therapy facility at the German Cancer Research Center. In: Burger G, Ebert HG (eds) Proceedings of the 2nd Symposium on Neutron Dosimetry in Biology and Medicine. Eur 5273 d-e-f, pp 445–452

Höver KH, Hesse BM, Rhein B, Lorenz WJ (1990) Verification of neutron dose distribution obtained by irregular shaped fields and moving beam therapy, In: Marin P, Mandrillon P (eds): Proc. 2nd European Particle Accelerator Conference, vol 2, pp 26–28

IAEA (1987) Report 277: Absorbed dose determination in photon and electron beams. IAEA, Vienna

ICRU (1983) Microdosimetry. ICRU report 36, International Commission on Radiation Units and Measurements, Bethesda, Md

ICRU (1989) Clinical neutron dosimetry, part I: determination of absorbed dose in a patient treated by external beams of fast neutrons. ICRU report 45, International Commission on Radiation Units and Measurements, Bethesda, Md

ICRU (1994) Prescribing, recording and reporting photon beam therapy. ICRU report 50, International Commission on Radiation Units and Measurements, Bethesda, Md

Jacobs FW, Van Kleffens HJ (1986) The "ring" method: a semi-empirical method to calculate dose distributions of irregularly shaped photon beams. Radiother Oncol 7:363–369

Loncol T, Vynckier S, Wambersie A (1996) Thermoluminescence in proton and fast neutron beams: Radiat Prot Dosim 66:299–304

Maughan RL, Yudelev M, Warmelink C, Forman JD, Maruyama Y, Porter AT, Powers WE, Blosser EB, Blosser G, Blosser HG (1994) Facility for fast neutron therapy at the Harper hospital. In: Amaldi U, Larsson B, (eds) Hadontherapy in oncology. Elsevier, Amsterdam, pp 377–385

Maughan RL, Blosser GF, Blosser EB, Yudelev M, Forman JD, Blosser HG, Powers WE (1996) A multi-rod collimator for neutron therapy. Int, J Radiat Oncol Biol Phys 34:411–420

Mijnheer BJ, Broerse JJ (1979) Dose distribution of clinical fast neutron beams. In: Barendsen GW, Broerse JJ, Breur K (eds) High-LET radiations in clinical radiotherapy. Pergamon, Oxford, pp 109–115

Mijnheer BJ, Wootton P, Williams JR, Eenmaa J, Parnell CJ (1987) Uniformity in neutron dosimetry. Med Phys 14:1020–1026

Pihet P, Meulders JP, Octave-Prignot M, Vynckier S, Wambersie A (1982) Gamma contribution to the total absorbed dose in the d(50)-Be and p(65)-Be therapeutic neutron beams at "CYCLONE". J Eur Radiother 3:121–130

Pihet P, Menzel HG, Meulders JP, Wambersie A (1984) Microdosimetric characteristics of high energy neutron beams and assessment of quantities relevant for radiation protection. Radiat Prot Dosim 9:241–244

Pihet P, Menzel HG, Schmidt R, Beauduin M, Wambersie A (1990) Biological weighting function for RBE specification of neutron therapy beams. Intercomparison of 9 European centres. Radiat Prot Dosim 31:437–442

Rassow J, Vynckier S, Hehn G (1984) Treatment planning in neutron therapy. J Eur Radiother 5:206–222

Rassow J, Haverkamp U, Hess A, Höver KH, Jahn U, Kronholz H, Meissner P, Regel K, Schmidt R (1992) Review on the physical and technical status of fast neutron therapy in Germany. Radiat Prot Dosim 44:447–451

Richard F, Meulders JP, Vynckier S, Octave-Prignot M, Possoz A, Wambersie A (1981) Present status of the treatment planning for d(50)-Be neutron beams and mixed schedules at "CYCLONE". Strahlentherapie 77 Suppl:178–184

Schmidt R, Hess A (1982) Spectroscopic intercomparison at the German neutron therapy centres. Int J Radiat Oncol Biol Phys 8:1511–1515

Schmidt R, Hess A, (1988) Component evaluation of event size spectra for a clinical 14-MeV neutron beam. Med Phys 15:343–347

Schmidt R, Thom E (1990) Neutron dose planning with the MEVAPLAN system. Strahlenther Onkol 166:301–305

Schmidt R, Schiemann T, Höhne KH, Hübener KH (1992) Three-dimensional treatment planning for fast neutrons. In: Breit A (ed) Advanced radiation therapy tumour response monitoring and treatment planning. Springer, Berlin Heidelberg New York, pp 643–648

Schmidt R, Schiemann T, Schlegel W, Höhne KH, Hübener KH (1994) Consideration of time-dose-patterns in 3D treatment planning, an approach toward 4D treatment planning. Strahlenther Onkol 170:292–301

Schraube H, Morhardt A, Grünauer F (1975) Neutron and gamma radiation field of a deuterium gas target at a compact cyclotron. In: Burger G, Ebert HG (eds) Proceedings of the 2nd Symposium on Neutron Dosimetry in Biology and Medicine, pp EUR 5273 d-e-f, 979–999

Vynçkier S, Octave-Prignot M, Richard F, Wambersie A (1981) Corrections for oblique incidence and wedges in a therapeutic d(50)+Be neutron beam. In: Schraube H, Burger G (eds) Proceedings of the 4th Symposium on Neutron Dosimetry, CEC, EUR 7448, vol 2, pp 129–140
Vynckier S, Pihet P, Octave-Prignot M, Meulders JP, Wambersie A (1982) Comparison of neutron therapy beams produced by 50 MeV deuterons and 65 MeV protons on beryllium. Acta Radiol Oncol 21:281–287
Vynckier S, Pihet P, Flemal JM, Meulders JP, Wambersie A (1983) Improvement of a p(65)+Be neutron beam for therapy at Cyclone, Louvain-la-Neuve. Phys Med Biol 28:685–691
Vynckier S, Mijnheer BJ, Rassow J (1985) Scaling relative neutron depth-dose distributions from one phantom material to another: a comparison of experimental and theoretical results. Phys Med Biol 30:55–66
Vynckier S, Pihet P, Marquebreucq S, Wambersie A (1987) The gamma component in clinical neutron beams of different energy. J Radiol 60:314
Vynckier S, Bonnett DE, Jones DTL (1991) Code of practice for clinical proton dosimetry. Radiother Oncol 20:53–63
Vynckier S, Patoul de N, Denis JM, Wambersie A (1993) Dosimetry of p(65)+Be neutron beams, produced by means of a multileaf collimator at Louvain-la-Neuve. Extended abstract book of the Workshop of the EORTC Heavy Particle Therapy Group and the European Clinical Heavy Particle Dosimetry Group ECHED, March 11–13, Brussels
Vynckier S, Bonnett DE, Jones DTL (1994) A supplement to the code of practice for clinical proton dosimetry. Radiother Oncol 32:174–179
Watermann FM, Kuchnir FT, Skaggs LS, Henrdy GO, Tom JL (1978) Dosimetric properties of neutron beams from D-D reaction in the energy range 6.8 to 11.1 MeV. Phys Med Biol 23:397–404
Wrede DE (1972) Central axis tissue-air ratios as a function of area/perimeter at depth and their applicability of irregularly shaped fields. Phys Med Biol 17:548–554

Specification of Radiation Quality in Fast Neutron Therapy: Microdosimetric and Radiobiological Approach

J. Gueulette[1], H.G. Menzel[2], P. Pihet[3], and A. Wambersie[1]

[1] Catholic University of Louvain, Radiobiology and Radioprotection Laboratory, 54, Avenue Hippocrate, B-1200 Brussels, Belgium
[2] European Economic Communities (EEC), Radiation Protection Research Action, DG XII/F6, Rue de la Loi, 200, B-1049 Brussels, Belgium
[3] CEA, Institut de Protection et de Sûreté Nucléaire, Département de Protection de la Santé de l'Homme et de Dosimétrie, Service de Dosimétrie, IPSN, BP No 6, F-92265 Fontenay-aux-Roses, France

Abstract

Specification of radiation quality is an important issue in fast neutron therapy since the biological effectiveness of the beams varies to a large extent with neutron energy. It must meet specific criteria, mainly derived from the accuracy requirement for absorbed dose delivery. A first approach to this problem consists in identifying physical parameters that can be related to Relative Biological Effectiveness (RBE) and which describe the beam production technique (e.g. neutron-producing reaction, p+Be or d+Be, energy of the incident particle). A second is based on microdosimetry, which provides a description of the secondary radiation components to which the biological consequences of irradiations are more directly correlated. A third approach consists in experimental RBE determinations in reference conditions: intestinal crypt regeneration in mice after irradiation to the whole body with single doses is proposed as a standard biological system for radiobiological calibrations of clinical fast neutron beams. Dosimetric, microdosimetric and radiobiological intercomparisons are encouraged since they provide a homogeneous set of data which facilitate the exchange of clinical information. They also constitute a basis for the clinical RBE approach and an overall check of the irradiation procedure. Therefore they should be recommended in every non-conventional radiation therapy facility.

Introduction: Importance of the Problem

Specification of radiation quality is an important issue in particle therapy and especially in fast neutron therapy since the biological effectiveness of the beams varies to a large extent with neutron energy. The RBEs of neutron beams are significantly higher than unity and vary significantly with energy and mode of production of the beams. Therefore, radiation quality must be specified in a quantitative and unambiguous way.

The introduction of neutron therapy has led to a large number of radiation quality investigations using biological and physical approaches. Radia-

Recent Results in Cancer Research, Vol. 150

tion quality specifications in radiation therapy must meet specific criteria. The main criterion is derived from the accuracy requirement for absorbed dose delivery. Any weighting factor to be evaluated from biological and physical experiments and to be applied to the absorbed dose must fulfill comparable accuracy requirements (Mijnheer et al. 1987). The validity of the factors can be restricted to the range of neutron energies used in therapy. The weighting factors would have to be used to support the establishment of reference RBEs with increased reliability and to enable RBE variations of clinical relevance at a given facility to be taken into account.

As will be discussed in "Experimental Determination of RBEs" below, radiobiological experiments in neutron therapy beams have been used to determine RBE values and RBE ratios. Such experiments were aimed at providing the basis for the reference RBE approach. Radiobiological investigations are absolutely necessary for the determination of the reference RBE and are required in any attempt to find a suitable radiation quality specification. However, their usefulness is limited by the inherently large uncertainties and unavoidable experimental difficulties. There are two alternative approaches (Kellerer and Rossi 1972; Menzel et al. 1990). The first consists of identifying physical parameters describing the radiation field that can be related to RBE. The second is based on microdosimetry and consists of studying the distribution of energy deposition by secondary charged particles put in motion by the neutrons, to which the biological properties of the beams are more directly correlated.

Relative Biological Effectiveness (RBE)

Historical Review

In radiation therapy with low-linear energy transfer (LET) radiations, the biological effect can be predicted from the dose level and the dose distribution. The specification of the type of radiation (gamma-rays, X-rays or electrons) and of the energy is of little importance since there is little RBE variation between gamma-rays (γ-rays), X-rays and electron beams from a few megaelectron volts to a few tens of megaelectron volts, at least for the effects relevant in radiation therapy. Therefore dosimetry alone and dosimetric intercomparison are needed.

In the past, however (1950 s–1960 s), when 200-kV X-rays (orthovoltage) were progressively replaced by cobalt-60 γ-rays (and electrons), a dose weighting factor of 0.85 was introduced to take into account the fact that, at equal doses, the 200-kV X-rays are somewhat more effective than high-energy radiations. The concept of Relative Biological Effectiveness (RBE) was recommended by the ICRP and the ICRU in 1963, then defined and discussed again in the International Commission on Radiation Units and Measurements (ICRU) Report 30 (1979). The conversion factor of 0.85 mentioned above was considered to be representative of the RBE of 200-kV X-rays relative to cobalt-60 γ-rays.

Later on, it became more and more obvious that the RBE value could be different for different systems and it has been shown that the RBE of 200-kV X-rays relative to cobalt-60 γ-rays could reach a value of 2 and more for some biological criteria, in particular at low doses, i.e. in the dose range which is more relevant in radiation protection (ICRU 1986). However, for the dose ranges and the effects relevant in radiation therapy, a conversion factor of 0.85 was appropriate to take into account the radiation quality differences between 200-kV X-rays and high-energy γ-rays or electrons (ICRU 1978).

The RBE issue is much more important in fast neutron therapy. Indeed, RBE values of neutrons (relative to γ-rays) ranging from about 1.5 to 5 were observed, depending on neutron energy, dose and biological effect. As a consequence, a single conversion factor could no longer be used (a single "RBE value" could not be assumed) to take into account these large RBE differences.

The RBE values for protons are small (1.0–1.2) and thus the RBE is of less relevance in proton therapy than for neutron therapy. For therapy using heavy ions, the RBE values are the highest (2–10) and exhibit the largest variations, especially in the distal region of the spread out Bragg peak (SOBP) where the LET of the heavy ion has its maximum.

Definition of RBE

When comparing two radiation qualities, one identifies the test radiation and the reference radiation. If D_{test} and D_{ref} are the observed dose necessary to reach a given biological effect for the test radiation and the reference radiation, respectively, the Relative Biological Effectiveness, RBE, of the test radiation, R_{test}, relative to the reference radiation, R_{ref}, is given by:

$$RBE_{test/ref} = D_{ref}/D_{test}$$

An RBE value is the result of an experiment and is thus associated with an experimental uncertainty. The RBE is a clear, unambiguous and well-defined concept. However, the biological system, type and level of effect, the dose and the experimental conditions, in which a given RBE value has been obtained, have to be specified.

Strictly speaking, the commonly used jargon that a certain radiation, e.g., fast neutrons from a given nuclear reaction, have a certain RBE is fundamentally incorrect and misleading. RBE values cited in this way are usually not the result of a given experiment but are judgments based on experimental RBEs and sometimes clinical experience. Although this practice is widespread and reflects some convenience of procedure for a complex issue, it has to be stressed very clearly that the use of such comparative values for different types of radiations should not be confused with RBE values. When comparing two radiation qualities there is not a single but a large number of RBE values, i.e. one RBE value for every set of system, effect and experimental conditions, (with the corresponding confidence intervals). Since the beginning of fast neutron therapy and even, before, a large number of determinations have been made. In some centres, comprehensive sets of RBE measurements were performed, e.g. the full RBE/dose relationships for different normal tissues and different types of tumours.

Some general conclusions can be derived from the available experimental data:

- Large differences in RBE values are observed (ranging from about 2 to 5).
- The neutron RBE increases with decreasing dose.
- The neutron RBE values are larger for late responding tissues than for early responding tissues. In other words, the neutron RBEs are higher for late complications than for early tolerance.
- The RBE also varies to a large extent with neutron energy, in the energy range used in therapy (about 40% for some systems).

Reference RBE

Due to the wide variation of RBE with the biological system and criterion, dose and experimental conditions, it is necessary to select reference conditions for RBE specification when transferring or exchanging clinical information. These reference conditions should be as indicative as possible for the clinical situations. The following conditions appear to be relevant:

- Dose level: 2 Gy (photon equivalent) per fraction
- Biological system: a system and endpoint representative of the RBE for average or overall late tolerance of normal tissues.

The RBE defined in such reference conditions could be called reference RBE ($_{ref}RBE_{n/\gamma}$).

The reference RBE is a radiobiological approach and, in principle, there should be only one reference RBE value for a given neutron beam. This implies that a single RBE value can be defined for an "overall" or "average" late tolerance for the normal tissues in patients. The fact that the alpha/beta ratios for late tolerance of different normal tissues are similar support this approach.

Most of the RBE determinations performed during the radiobiological pre-therapeutic experiments and during the radiobiological intercomparisons were using biological systems chosen since they were suitable for the type of experiment and not necessarily since they were representative: well codified, reliable, easy to transport, providing reproducible results (e.g. mammalian cells in vitro, *Vicia faba*, intestinal crypt cell systems (Gueulette et al. 1986; Hall et al. 1992). A general and formal agreement does not exist as yet.

Clinical RBE

In contrast to the reference RBE, the clinical RBE is a clinical and operational concept. The term "clinical RBE" has been used especially in the

United States and is understood as a ratio of the absorbed dose which would have been given in a photon treatment and the neutron dose which is actually prescribed at a given neutron therapy facility (and for a given tumour localisation) (Wambersie and Battermann 1987; Wambersie and Menzel 1993). Dose prescription and treatment planning also depend on other factors, such as beam penetration or, in general, geometric factors. If the physical selectivity of the neutron beams used is significantly inferior to that achieved with photons, the neutron dose obtained by simply dividing the photon dose by the reference RBE often cannot be delivered because the expected dose to the normal tissues at risk would exceed the tolerance level. Therefore, different and more complex beam arrangements have to be investigated for neutrons. If the dose to the normal tissues at risk are still too high, a dose reduction factor taking into account the inferior physical selectivity of the neutrons has to be applied.

At a given facility, there could be several clinical RBEs depending on the tumour site, especially when poorly penetrating beams are used. However, at modern facilities, the physical selectivity of neutrons is comparable to that of photons, and the clinical RBE tends to become close to the reference RBE (Wambersie 1992). The "clinical RBE" can be defined as the ratio of the photon and neutron doses which are actually delivered to the patient in a given neutron therapy facility and in a given therapeutic protocol. It is the quantity the radiotherapist has to select when prescribing the irradiation (e.g. in terms of monitor units). Although it is a dose ratio for two radiation qualities, it is not in the strict sense an RBE: it is the reference RBE (as defined above) weighted by empirical account of the physical selectivity (when needed) and the clinical experience (when available). It implies a value based on the judgment of the radiotherapist.

Physical Beam Parameters

A characterisation of the radiation quality can be derived from the description of beam production technique. For fast neutrons, this includes the neutron-producing reaction (p+Be or d+Be), the energy of the incident particles, the target composition and geometry, and other geometrical factors (e.g. collimation system, additional polyethylene filter). A first indication of the neutron energy can be derived from these parameters. The neutron energy spectrum can also be measured by spectroscopic methods. As in current practice, single parameters appear more useful than complete spectral information; the mean neutron energy or related parameters, such as the half-value thickness (HVT), have also been used (see, e.g. Fig. 11; Gueulette et al. 1996).

In the beginning of fast neutron therapy, it was common practice to describe the neutron beam in terms of the "neutron component" and "gamma component". Also at that time (and for many applications), an RBE value of 3 was assigned to the neutron component and an RBE value of 1 to the gamma component. The development of the microdosimetric programmes has shown that this approach is indeed an oversimplification and that several types of secondary charged particles must be taken into account (see, e.g. Fig. 2; Pihet et al. 1990).

Microdosimetric Approach to Specifying the Radiation Quality in Fast Neutron Therapy

Microdosimetric Data

Specification of the neutron energy, in terms of spectrum or average value, is primary information. Neutrons deposit energy ultimately through secondary charged particles whose type and energy distributions critically depend on the shape of the neutron spectrum and on the type of material with which they interact. The biological consequences of irradiations by neutron beams are therefore more directly correlated to the secondary radiation spectra and only indirectly to the neutron energy. However, the spectra of secondary charged particles are not an easily accessible piece of information, especially for high-energy neutron beams. The measurement of microdosimetric spectra offers, at least for the restricted field of neutron therapy, a sufficiently accurate description of the secondary radiation components and provides comprehensive and detailed information on radiation quality.

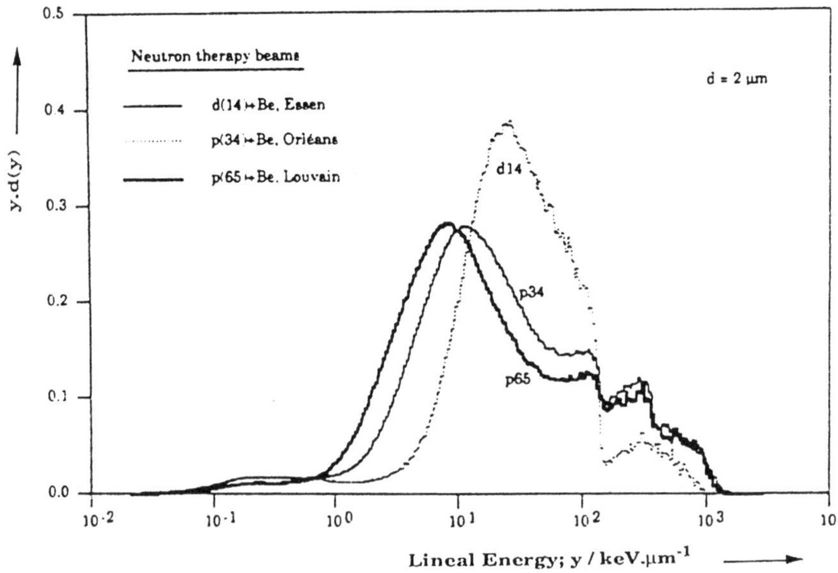

Fig. 1. Microdosimetric intercomparison of neutron therapy beams at European facilities. Comparison of the linear energy spectra obtained for low-energy neutron beam and for two beams of the new generation (high-energy proton on Be targets) neutron beams is shown. (From Pihet 1989)

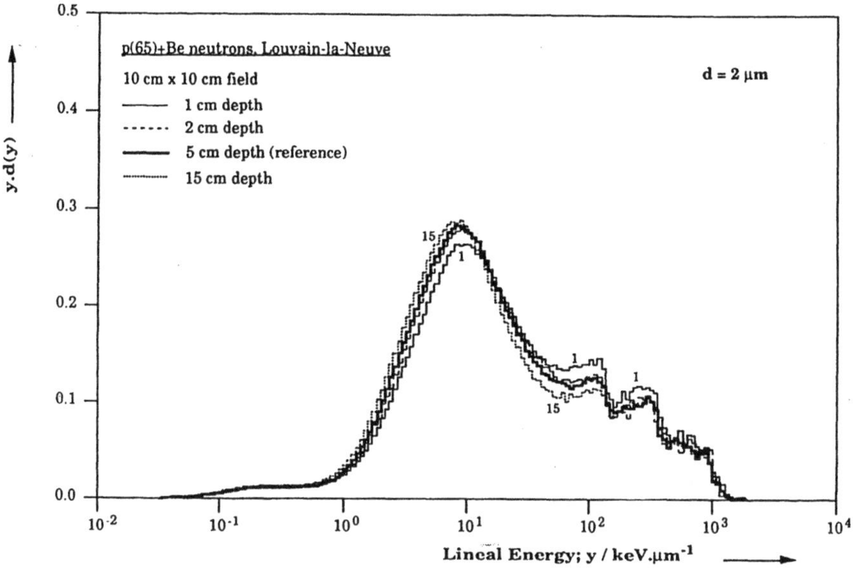

Fig. 2. Comparison of microdosimetric spectra measured at different depths in a phantom on the axis of the p(65)+Be neutron therapy beam used at Louvain-la-Neuve (the highest energy used in neutron therapy). Four peaks can be identified in the p(65)+Be neutron spectra: the first is at 8 keV μm^{-1} and corresponds to high-energy protons, the second is at 100 keV μm^{-1} and corresponds to low-energy protons, the third is at 300 keV μm^{-1} and is due to α-particles and the last at 700 MeV μm^{-1} and is due to recoil nuclei. The (rather flat) differences between the spectra measured at different depths are relatively small and no significant RBE variation has been detected. (From Pihet 1989)

Several microdosimetric studies have been carried out by different groups since the early 1970s in close connection with the development of neutron therapy (Wilson and Field 1970; Pihet et al. 1988). These studies were aimed at the investigation of changes in radiation quality between different neutron therapy beams and for a given beam between different irradiation conditions. Figures 1–4 show typical examples of the results obtained from these investigations: comparison of the dose distribution in lineal energy measured under similar conditions for neutron beams of different energies at various therapy facilities (Fig. 1); comparison of microdosimetric spectra obtained in a neutron beam with a given energy at different depths in a phantom, on the axis of a standard 10×10 cm field (Fig. 2); comparison of microdosimetric spectra obtained in a neutron beam with a given energy, at the same depth in the phantom, on the beam axis, and for different field sizes (Fig. 3); comparison of microdosimetric spectra obtained in a neutron beam with a given energy, outside the geometrical beam, and by using collimators made of different materials (Fig. 4). These examples illustrate the potential of the microdosimetric approach to account implicitly for any factor related to the current clinical practice that may influence radiation quality.

Although the correlation between the shape of the lineal energy spectra and the quality of the neutron beams appears obvious for neutron beams

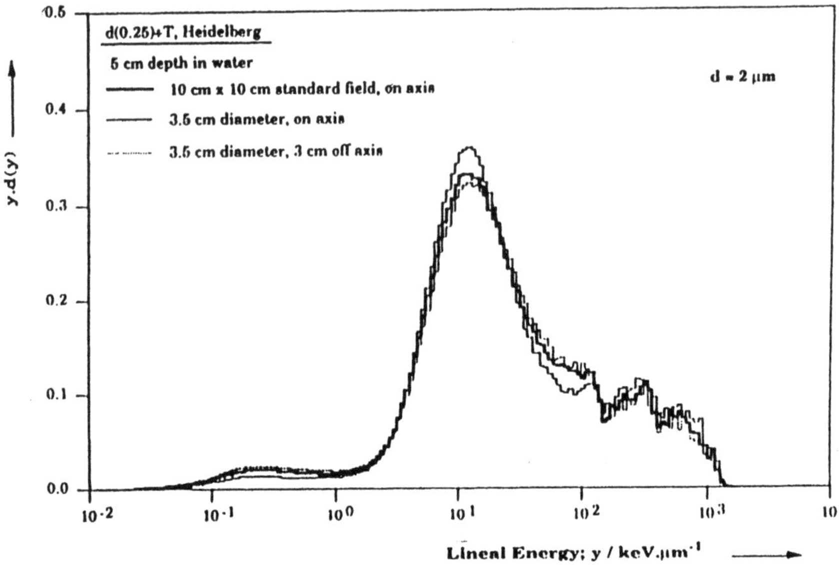

Fig. 3. Comparison of microdosimetric spectra measured for the 14 MeV d+T neutron beam in Heidelberg for different field sizes. (From Menzel et al. 1990)

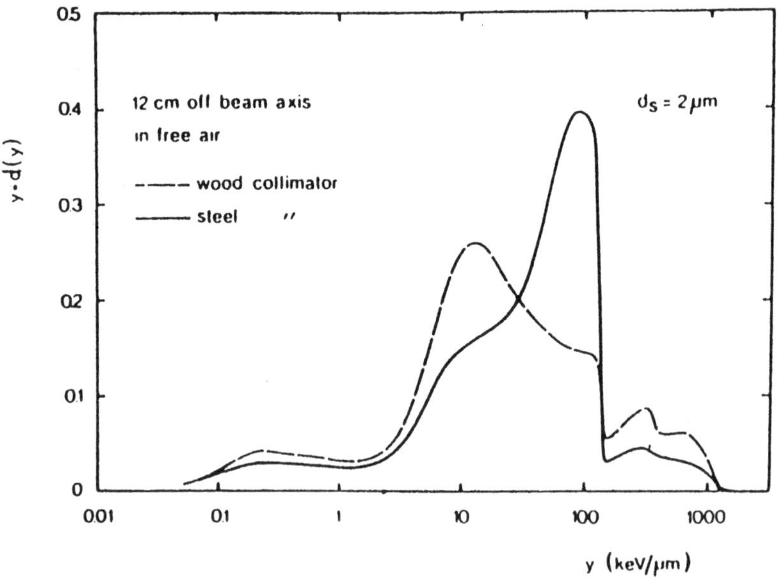

Fig. 4. Comparison of microdosimetric spectra measured for 14 MeV d+T neutrons (DKFZ Heidelberg) outside the geometrical beam for different types of collimators

with widely differing energies, the use of microdosimetric spectra for quantifying the variation of radiation quality between different neutron beams by simply using purely physical parameters, such as mean lineal energies, presents in general problems.

The lineal energy spectra for fast neutrons show large overlapping components of different types of particles (e^-, p, α, heavier ions). Variation of radiation quality between neutron beams of different energy is therefore reflected in competing changes in the different components of the spectra. RBE versus LET curves for many biological endpoints show a maximum, usually around 100 keV μm^{-1}. Consequently, the contribution of each lineal energy component to the average radiation quality for a given neutron beam critically must depend on both their contribution to the total absorbed dose and on their relative biological weight.

The quantification of radiation quality is particularly difficult if neutron beams with small differences in their neutron energy spectra are considered. Such small variation may lead to a significant change in intervals where RBE is high, i.e. around the peak of the RBE versus LET or y-curve. As an example, in Fig. 5, the microdosimetric spectra measured for two neutron beams produced at two different facilities by using the same reaction and energy are compared. The major difference occurs between 20 and 100 keV μm^{-1}, which is mainly due to the contribution of recoil protons with relatively low energy. This variation is physically explained by the different target thickness used in the two centres. Biologically, an RBE difference of 6% has been ob-

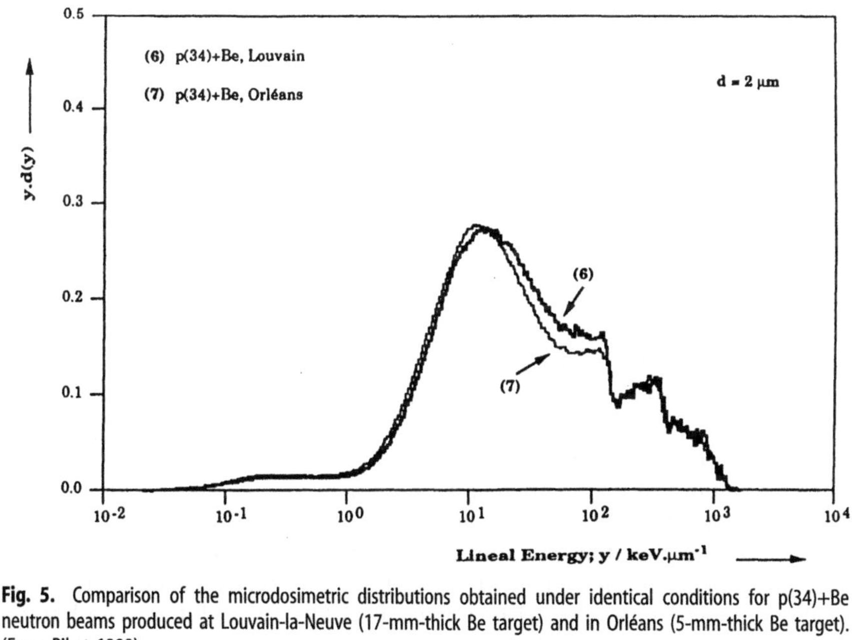

Fig. 5. Comparison of the microdosimetric distributions obtained under identical conditions for p(34)+Be neutron beams produced at Louvain-la-Neuve (17-mm-thick Be target) and in Orléans (5-mm-thick Be target). (From Pihet 1989)

served between these two neutron beams. This difference is not negligible considering the accuracy requirements in radiation therapy.

Another example is given by the comparison of microdosimetric spectra measured for the same neutron beam, e.g. [p(65)+Be] in Fig. 2 at different depths in a phantom. Here again the differences between the spectra are relatively small but they affect the lineal energy interval where the biological weight of the events is most important. Related differences in RBE, however, may be detected in biological experiments only if large efforts are made to achieve a sufficiently low level of experimental uncertainty.

In order to solve the problem of specifying radiation quality in neutron therapy, it would be useful to identify a parameter with a relative variation similar to that of RBE. It would then be necessary to prove that such a parameter could be determined with sufficiently low uncertainty.

Empirical Biological Weighting Function

The approach presented here was presented by Pihet (1989). It is based on the principle of using a biological weighting function applied to the dose distribution in lineal energy for a given radiation field in order to determine a single parameter that estimates its quality. This procedure is well known in microdosimetry and was applied, for example, in radiation protection for determining the mean quality factor for a given radiation field (Hartmann et al. 1981):

$$Q = \int \mathrm{q}(y) \times \mathrm{d}(y) \times \mathrm{d}y \tag{1}$$

Applying this principle to the field of neutron therapy, the problem of specifying the radiation quality for a neutron beam with a given energy compared with that of another neutron beam chosen as a reference may be solved by optimizing a weighting function $\mathrm{r}(y)$ so that the integral R:

$$R = \int \mathrm{r}(y) \times \mathrm{d}(y) \times \mathrm{d}y \tag{2}$$

reproduces the RBE ratio between the two neutron beams [13]. This approach only assumes a correlation between the RBE and the shape of its microdosimetric dose distribution. It does not require further assumption regarding the biophysical meaning of the energy actually deposited in the site.

The specification of radiation quality for neutron therapy beams requires that the parameter R is determined with an uncertainty of about 3%. The crucial problem therefore remains as to how accurately the weighting function $\mathrm{r}(y)$ can be optimized in order to fulfil this requirement. During the 1980 s, biological intercomparisons of neutron therapy facilities were limited most often to two neutron beams of different energies (Zywietz et al. 1982;

Bewley et al. 1989; Joiner and Field 1988). Later, systematic biological intercomparisons of neutron therapy beams in the energy range between [d(14)+Be] and [p(65)+Be] became available (Beauduin et al. 1989a). At the same time, a microdosimetric intercomparison carried out under the auspices of the EORTC (European Organization on Research and Treatment of Cancer) enabled the measurement of the microdosimetric characteristics for 14 different neutron beams including those used in the biological experiments (Pihet et al. 1989). By using the microdosimetric spectra and the RBE ratios determined for the same neutron beams as input data, the weighting function r(y) could be optimized numerically by an iterative procedure. This unfolding method has been applied several times in microdosimetry to evaluate empirically biological weighting functions (Varma and Bond 1982; Zaider and Brenner 1985). Assuming an initial guess function, the parameters of the function r(y) are optimized by successive iteration (Eq. 2) in order to match the calculated parameter type R and the experimental RBE ratio for each neutron beam. The main limitations of this approach are the energy range, the biological endpoint and the dose level for the RBE values used as input data. This calculation could be performed using the data for nine neutron beams of different energies ranging from [d(4)+Be] to [p(65)+Be] (Pihet et al. 1990). The optimized weighting function found by using two different series of RBE ratios is shown in Fig. 6. Their shape is similar to that of the RBE versus LET curves. The effect of folding the weighting function with

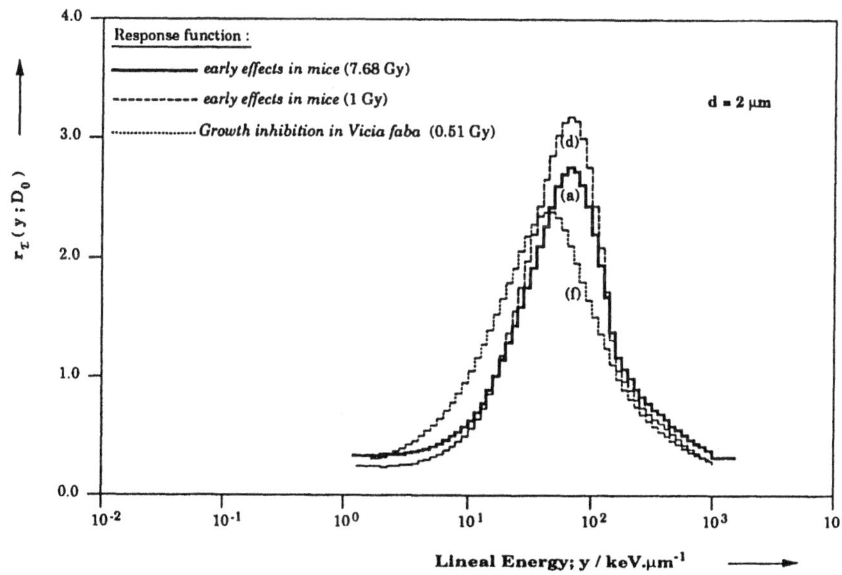

Fig. 6. Biological weighting functions obtained numerically by unfolding calculation using RBE ratios and microdosimetric distributions obtained for the same neutron beams as experimental input data. The functions depend on the biological system and the dose level corresponding to the input RBE value. (From Pihet 1989)

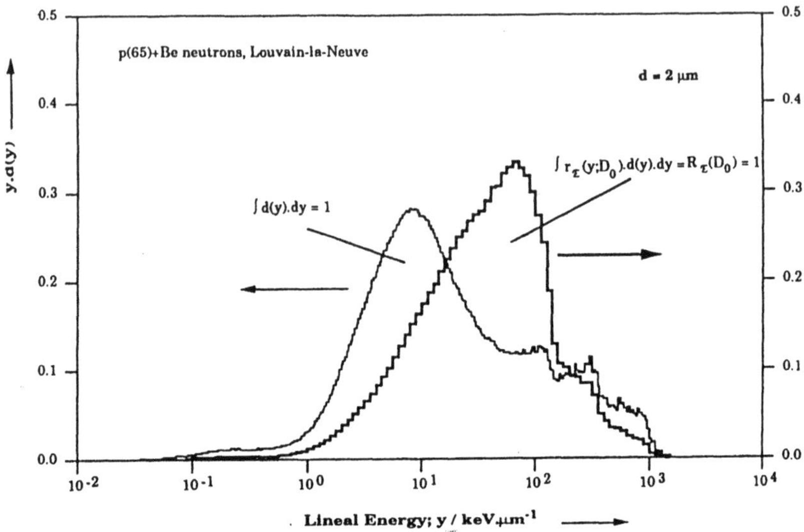

Fig. 7. Comparison of the dose distribution in linear energy d(y) for p(65)+Be neutrons and the corresponding weighted distribution r(y)d(y) (for early effects). The integral under the weighted curve gives an estimate of the RBE of the neutron beam. This integral is equal to 1 for p(65)+Be neutrons since this beam has been used as reference in the related biological experiments. (From Pihet 1989)

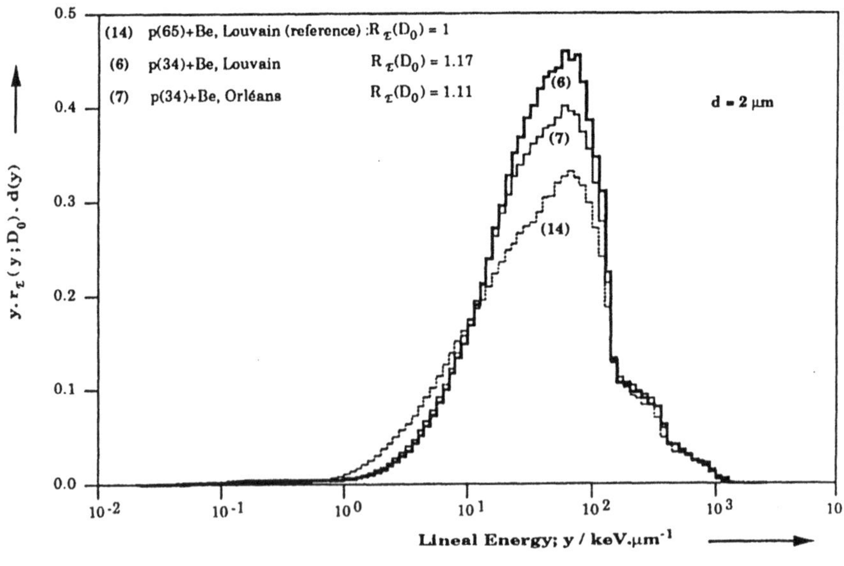

Fig. 8. Comparison of the weighted dose distribution (for early effects) in lineal energy obtained for p(34)+Be neutron beams produced at two different facilities and for the p(65)+Be neutron beam taken as a reference. The corresponding unweighted distributions were compared in Fig. 5 (From Pihet 1989)

the dose distribution in lineal energy is shown in Fig. 7 for [p(65)+Be] neutrons. The integral of the weighted dose distribution, *R*, gives an estimate of the RBE of the beam. In the case of [p(65)+Be] neutrons, this integral is equal to 1 as this neutron beam was taken as the reference radiation according to the biological experiments (RBE ratio = 1). The weighted dose distributions enable the identification of the secondary radiation components that are mainly responsible for the differences in RBE. This is illustrated in Fig. 8 in comparison to Fig. 5; the RBE ratio of [p(34)+Be] neutrons in Louvain-la-Neuve compared with [p(34)+Be] neutrons in Orléans was found to be about 1.06, which agrees well with the difference between the weighted spectra in the interval between 50 and 150 keV μm^{-1}.

Experimental Determinations of RBEs

Pretherapeutic and Preclinical Experiments

Before applying new types of radiations in cancer therapy, one can identify two types of radiobiological experiment: A first type of experiment (which will be called "pretherapeutic" experiments) aims to identify the rationale of the new radiation. For example, introduction of fast neutrons in therapy was based on the evidence of the hypoxic cancer cells and on a reduction of oxygen enhancement ratio (OER) with neutrons. Radiobiological experiments also showed that neutrons (or high-LET radiations) are specifically efficient against cells in some radiosensitive phases (e.g. S) of the mitotic cycle. They also provided information about the (optimal) way to apply these radiations. For example, it was shown that there is less cell repair with high-LET radiations, which should influence the fractionation scheme. Finally, the pretherapeutic experiments can also provide information about the clinical indications for the new type of radiation. For example, tumours with large hypoxic areas which reoxygenate slowly and well-differentiated slowly growing tumours are expected to be good indications for neutron therapy.

This first type of experiment is usually performed before the construction of the neutron therapy facility is decided. Actually, the construction of the facility is the consequence of the pretherapeutic radiobiological programme. When the new facility is completed and the beam ready for application, a second type of experiment (which will be called "preclinical" experiments) is performed. This type only has reference to RBE determinations and aims to clinically apply the neutrons under safe and optimal conditions. Preclinical experiments in neutron beams are particularly important as RBE values are high (2–5) and do vary largely with dose and biological effect. Moreover, RBEs depend significantly on both the energy and the mode of production of the beams and could vary (especially for the p+Be reaction) with depth. As a consequence, the determination of RBE values relevant to radiotherapy raises complex problems related to, e.g. the choice of the biological system and the irradiation conditions.

Respective Merits of Different Biological Systems for Preclinical Studies

Cell Survival In Vitro. Mammalian cells grown in monolayer have been widely used for RBE measurements since 1956 (Puck and Markus 1956). The improvement of biotechnology makes it possible to cultivate in vitro many mammalian cell lines. However, well-established cell lines, e.g. V79, EMT6, CHO, are particularly useful since they have a high plating efficiency (60%–80%) and are relatively easy to cultivate under standardised conditions. They also have a radiosensitivity and a repair capacity corresponding to the average normal or cancer cells.

The determination of cell survical curves is a priori suitable since both acute and late biological effects that account for radiotherapy can be interpreted from cell lethality (Gueulette et al. 1994). However, cell culture in vitro remains artificial and it is difficult to evaluate the extent to which the results obtained for a particular cell line could be extrapolated to the corresponding tissue. The choice of the cell line should thus be more dictated by its convenience (ease of culture and plating efficiency).

Vicia faba. Growth inhibition in *Vicia faba* bean roots is a well-documented vegetal system (Van Dam 1984) which has been, and still is used for RBE and OER determinations (Fig. 9). The endpoint (G10) is the relative growth of the central root over the 10-day period following the irradiation, the mean growth of unirradiated roots being taken as the reference. The sensitive part of the bean (target volume) to be irradiated at the prescribed dose is the root tip (much less than 1 cm long and less than 2 mm in diameter); the dose delivered to the other parts of the bean is of little importance for the growth of the root.

Vicia faba is suitable for preclinical studies since growth inhibition can be considered as a tissular effect related to the survival and the proliferating capacity of root meristem cells (Van Dam 1986). The bean roots are highly radiosensitive, so that single fraction irradiations make it possible to investigate small doses (0.5–2 Gy gamma equivalent) and to determine RBE values in a dose range relevant to radiotherapy. The main indication for using *Vicia faba*, however, is OER determinations, which are much easier to perform than in any in vitro cell system. In this respect, it is worth noting that OER, which depends on the radical formation, is the same for all biological systems (which is not the case for RBE).

Chromosome Aberrations in Vegetal Systems. Determination of radio-induced chromosome aberrations in vegetal roots is another system developed for RBE determination (Fig. 10). While *Vicia faba* is most popular (Hall and Cavanagh 1967), in our laboratory, for many years we have used *Allium cepa* onion roots (variety *"jaune paille des vertus"*) following Markus and Kolliopoulou (1965). *Allium cepa* has relatively few chromosomes ($2N=16$), has large-size chromosomes ($\sim$7 mm) and exhibits frequent and regular mitoses.

Chromosome aberrations are interesting in so far as they are correlated with cell death and allow the investigation of a (very) low dose range (0.2–

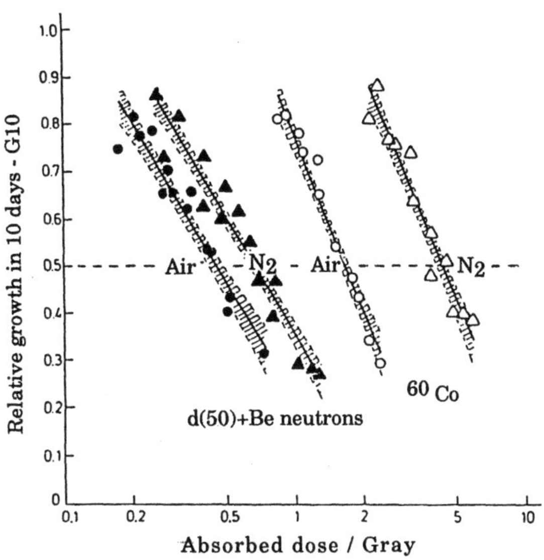

Fig. 9. Relative growth of *Vicia faba* root tips in 10 days as a function of dose (logarithmic scale) after irradiation under aerobic and hypoxic conditions with d(50)+Be neutrons and cobalt-60 γ-rays (*shaded areas* correspond to the 95% confidence interval). The slopes obtained under aerobic and hypoxic conditions do not show a significant difference for either radiation quality (OER does not depend significantly on dose). On the other hand, the slopes of the dose-effect relations for neutrons and γ-rays are different, which implies an increasing RBE with decreasing dose. (From Van Dam 1984)

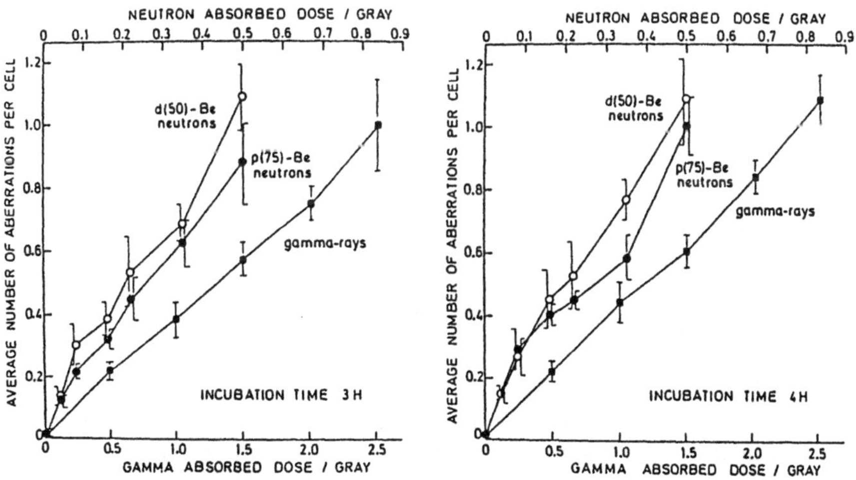

Fig. 10. Average number of aberrations per cell in anaphase and telophase as a function of absorbed dose for d(50)+Be, p(75)+Be neutrons (*upper abscissa*) and cobalt-60 gamma-rays (*lower abscissa*). The incubation times between irradiation and fixation are 3 h (left-hand side) and 4 h (right-hand side). No significant RBE differences were observed for longer incubation times up to 7 h. (From Beauduin et al. 1992)

2.0 Gy gamma equivalent). On the other hand, chromosome aberrations are known to provide larger RBE variations with LET than mammalian cells (Barendsen 1978), which brings to the system the ability to distinguish small changes in radiation quality. In addition, due to the small size of their sensitive part, *Allium cepa* and (to a smaller extent) *Vicia faba* are suitable for RBE determinations in very small irradiation fields.

Intestinal Crypt Cell Regeneration in Mice. This in vivo system in mice, introduced by Withers and Elkind (1970), was developed to derive the cell survival curve for intestinal stem cells. Taken as a model for the evaluation of the early tolerance of (normal) tissues, this system was widely used to study the influence of fractionation, dose rate, effect of radiosensitisers and also to determine RBE under different conditions (see, e.g. Fig. 13).

Intestinal crypt regeneration in mice deserves a privileged place in preclinical studies in so far as the experimental irradiations may be carried out in the same spirit as clinical irradiations where different problems related to, e.g. the definition of the target volume, point of dose and RBE specification, have to be faced. Owing to its importance, especially for radiobiological intercomparisons (see below), the full procedure is given in the Appendix.

Mouse Lung Tolerance. Lung tolerance in mice, assessed by $LD_{50}/180$ (dose corresponding to 50% animal survival at 180 days), has been taken as a model for the evaluation of the late tolerance of (normal) tissues (Grégoire et al. 1993). This system is one of those allowing the determination of a reference RBE, which implies the use of fractionated irradiations. Ten fractions separated by a time interval of 12 h appears to constitute a good compromise between the necessity of allowing complete repair of the sublethal damage and of avoiding "slow repair" that could occur when the irradiations are spread over a long time (Field et al. 1986).

Radiobiological Intercomparisons

Radiobiological intercomparisons are the obligatory counterpart of dosimetric intercomparisons when radiation beams are presumed to differ in RBE. They aim at facilitating the comparison of the clinical results and the pooling of experience. On the other hand, they should make it possible to transfer particular radiobiological or clinical information to other facilities, allowing them to bypass certain points of the studies.

Radiobiological intercomparisons have become compulsory in neutron therapy since the RBE of the different clinical beams is largely variable. Indeed, neutrons are produced by bombarding a beryllium target with protons or deuterons and have different "effective" energies depending on the nuclear reaction and the energy of the incident particle. Moreover, the energy (and thus the RBE) of a given beam could further vary according to several parameters peculiar to each facility. For example, the collimation system (type and composition of the collimator), the thickness of the beryllium target or the filtration of the beam could influence the RBE for more than 5% (i.e. beyond the dose accuracy needed in radiotherapy). If the changes in the beam quality from one neutron therapy facility to another could be predicted from the information discussed above (physical beam parameters, microdosimetry) and if the magnitude of the corresponding RBE variations could be calculated and predicted, their actual value would still have to be measured.

Since the beginning of fast neutron therapy, several systematic intercomparisons have been performed, using biological systems as different as: vegetal systems (growth inhibition of *Vicia faba* root (Beauduin et al. 1986; Van Dam and Wambersie 1981); chromosome aberration in *Allium cepa* (Beauduin et al. 1992); in vitro systems (V79 and CHO (Guichard et al. 1980) cell survival); in vivo systems [intestinal crypt cells survival in mice, LD_{50} in mice after thoracic (Grégoire et al. 1993), or intestinal (Wambersie et al. 1978) irradiation, ethylendiaminetetraacetate (EDTA) clearance in mouse kidney (Joiner 1989), skin reaction in mouse (Joiner et al. 1983) and weight loss and DNA content in mouse testis (Geraci et al. 1997)]. However, when the beam quality (i.e. the LET) differs only by a small amount, the RBE differences related to the biological system are negligible. Therefore, the choice of a biological system for intercomparisons should be more especially dictated by its convenience, portability and reproducibility (Hall and Kellerer 1979). Crypt cell regeneration in mice (as described in the Appendix) meets these requirements since it is particularly independent of the environmental conditions, easy to handle and capable of a high degree of precision due to its steep dose-effect relationship. It was used by our team for the intercomparison of neutron beams produced at different facilities worldwide (Gueulette et al. 1996).

The variation in the RBE of these beams is presented in Fig. 11, which indicates that the RBE ratio between the beams with the lowest [d(14)+Be] and the highest energy [p(75)+Be] rises above 1.5. These RBE values should be used only for the characterisation of the neutron beam quality and should not be used as the "clinical RBE" (or clinical RBE ratios), i.e. to derive the prescribed dose to the patient. Indeed, they were obtained for an early effect and with a single large fraction, and it is well known that they are smaller than RBE values (or RBE ratios) obtained for late effects and for small doses per fraction which account for the determination of the reference or the clinical RBE.

In addition to useful radiobiological information, direct radiobiological intercomparisons provide an overall check on the dosimetric and irradiation procedures. They should be performed before starting any clinical programme and should be associated with dosimetric intercomparisons and, when possible, which microdosimetric measurements. Therefore, it is important that the experiments be performed in collaboration with the radiobiologists and dosimetrists from each facility and according to a strict procedure.

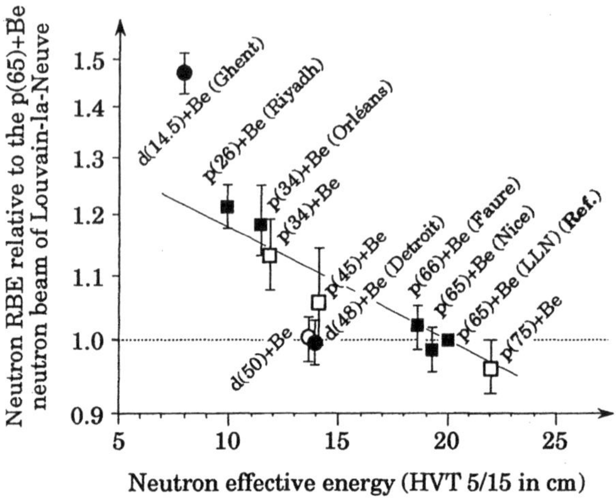

Fig. 11. Variation in the RBE of different neutron beams as a function of energy. The p(65)+Be neutron beam at Louvain-la-Neuve has been taken as the reference; the data represent the RBE for intestinal crypt regeneration after single fraction irradiation at 5 cm in depth in a water phantom (2 cm for Detroit and Nice). The effective energy of the beams is expressed by HVT 5/15 measured under reference conditions (Gueulette et al. 1996). The *closed symbols* correspond to neutron beams of the different facilities visited; the *open symbols* correspond to neutron beams produced at the variable-energy cyclotron at Louvain-la-Neuve. *Squares and circles* correspond to neutron beams produced from proton and deuterons, respectively. *Error bars* indicate the 95% confidence interval. A *straight line* was fitted through the points corresponding to neutron beams produced from protons (*closed and open squares*). (From Gueulette et al. 1996)

Appendix: Procedure for Crypt Cell Regeneration Assay

Principle and Biological Endpoints. The intestinal mucosa may be described as a two-compartment tissue. The first compartment contains the stem cells, which are able to proliferate and are located in the lower part of the crypts. The second compartment contains differentiated and functional cells which are no longer able to divide and which line the "villi" (Fig. 12). Only the stem cells (crypt cells) are radiosensitive and it is assumed that a single surviving stem cell can regenerate a crypt. Therefore, according to the Poisson law, the number of observed regenerated crypts and the number of surviving stem cells can be considered to be equal when less than about 70 regenerated cells are counted per circumference; this corresponds to a dose range of 12–20 Gy (gamma equivalent).

Biological Procedure. The conditions described hereafter are those which have been used in our laboratory for many years (Gueulette et al. 1996). However, modifications in the experimental protocol are allowed. Female NMRI, Balb/c or C57 black, 11–13 weeks old, were used. These three strains exhibit the same radiosensitivity. The mice are housed in plastic cages (six to eight mice per cage) and randomised with respect to different parameters such as dose level, radiation quality (neutrons or reference cobalt-60 γ-rays) and, depending on the type of experiment, fractionation scheme. Food and water is pro-

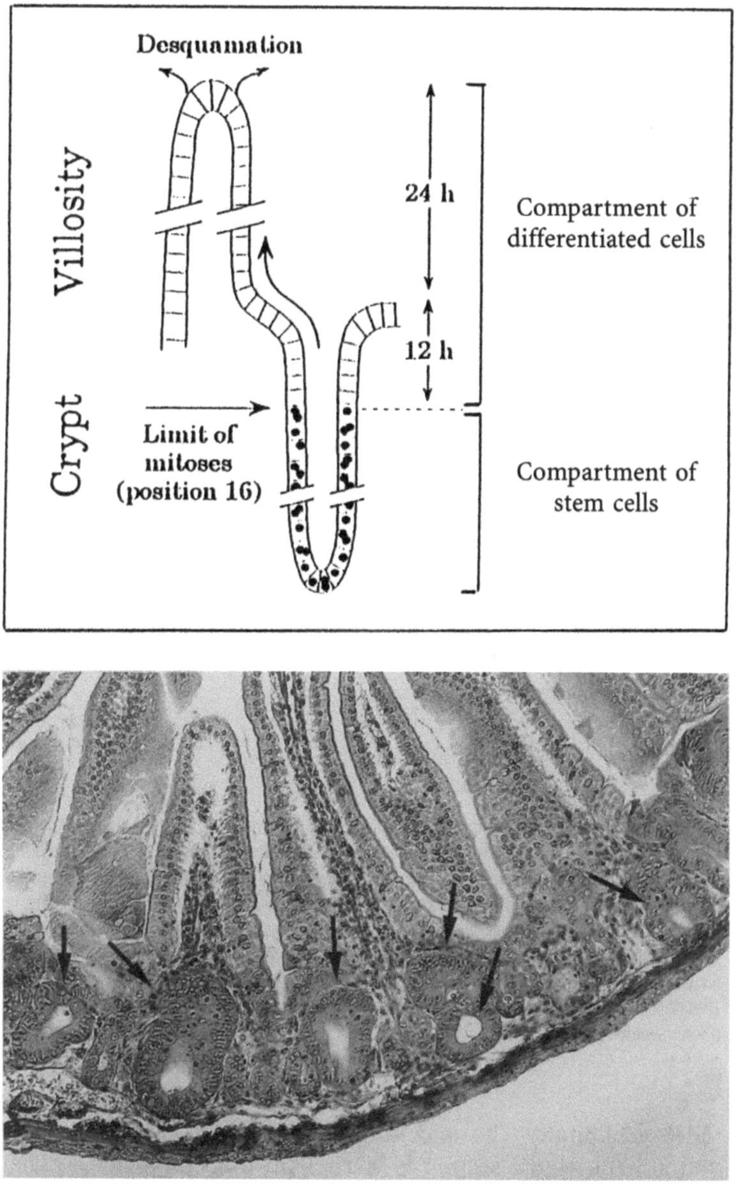

a

b

Fig. 12. *Upper*: Schematic view of a histological section of the intestinal mucosa, which would go both through the axis of a crypt and the axis of a villosity. Part of the 32 crypt cells "piled up" along the longitudinal axis have been drawn. The superior limit of divisions (position 16) practically demarcates the compartments of stem cells and differentiated cells. The direction and the time of the migration of differential cells is indicated.

Lower: Transversal section of the intestinal mucosa, 3 days after irradiation by 12.6 Gy (gamma equivalent). The length of the villosities is reduced and the crypts have practically disappeared. Only a few piles of apparently viable and nucleated cells are visible, which correspond to the regenerated crypts (see *arrows*). They result from the multiplication of one cell which kept its proliferating capacity. (From Gueulette 1995)

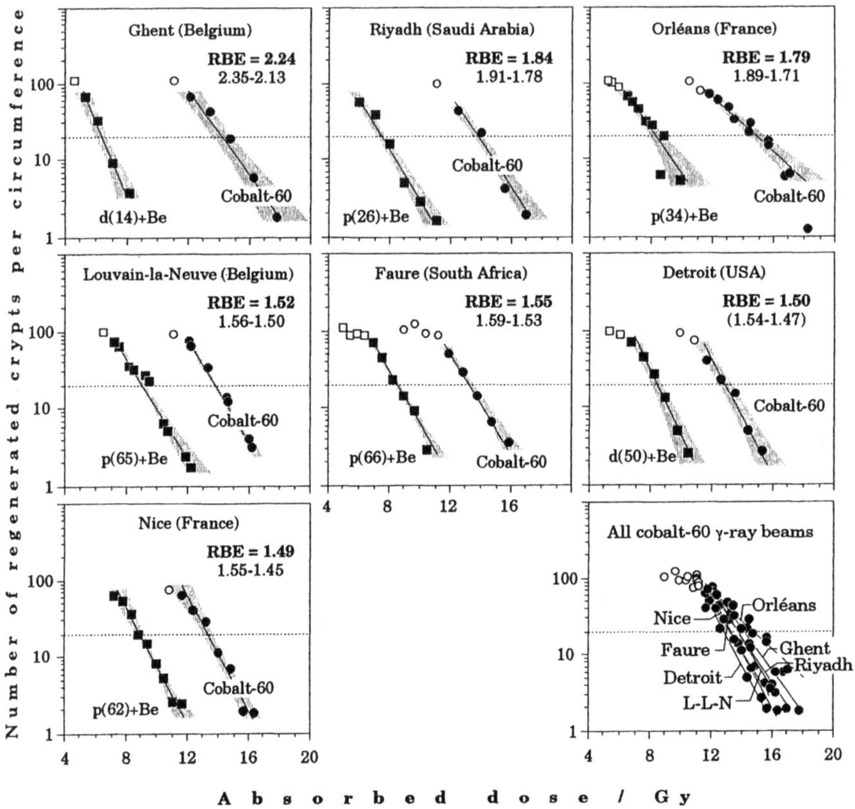

Fig. 13. Dose-effect relationships for intestinal crypt regeneration in mice after irradiation by neutrons and cobalt-60 at different neutron therapy facilities. Exponential curves were fitted through the experimental points. The open points were not taken into account: they correspond to ≥70 crypts per circumference (see text). *Grey areas* correspond to the confidence intervals of the slopes of the curves. The indicated RBEs and their 95% confidence intervals were calculated for 20 regenerated crypts per circumference; they are plotted as a function of the beam energy in Fig. 11 ("visited facilities", *closed symbols*). The dose-effect relationships for cobalt-60 gamma-rays obtained at the different facilities are put together on the *lower right-hand panel.* The doses corresponding to 20 regenerated crypts per circumference do not differ by more than ± 7%. (From Gueulette et al. 1996)

vided ad libitum. The mice are sacrificed 84 h after the end of irradiation and a 3-cm jejunal section is taken 1 cm from the pylorus. The samples are immediately fixed in "Boin Hollande" (Gabe 1968). They are then embedded in paraffin, transversely sliced (5 µm sections) and finally stained according to the classical trichrome technique.

Irradiation. The mice are not anaesthetised and are kept in an elongated position using wooden clothespegs (Wambersie et al. 1978). They are located in a Perspex mounting where their abdomen is squeezed between two thin plates 1.5 cm apart from one another. The Perspex mounting fits in a polystyrene phantom so that the mid-abdomen is situated accurately at the assigned position in the beam.

Statistical Analysis. The basic experiment compares the dose-effect relationships obtained after single fraction irradiation (Fig. 13). In this case, an exponential relation can usually be assumed for the number of regenerated crypts as a function of dose and a straight line (in log/linear coordinates) can be fitted through the experimental points. This results from the fact that the doses explored in the crypt regeneration assay are rather high and correspond to the (exponential) distal region of the cell survival curve. However, the identity between the number of regenerated crypts and the number of surviving stem cells tends progressively to disappear at small doses for which high numbers of regenerated crypts are counted. Therefore, the points corresponding to more than about 70 regenerated crypts per circumference are not taken into account for computation: at this level, the discrepancy between the number of regenerated crypts and surviving stem cells reaches about 7% (for an average of 140 crypts per circumference). Most often the RBE values are determined at the level of 10 or 20 crypts per circumference. With six to eight mice per dose level and eight to ten dose levels for the compared dose-effect relations, the accuracy on the RBE value is usually ±3.5% for the biological procedure (independently of the dosimetric uncertainties for the compared beams).

References

Barendsen GW (1978) RBE-LET relations for induction of reproductive death and chromosome aberrations in mammalian cells. In: Booz J, Ebert HG (eds) 6th Symposium on Microdosimetry, Brussels. Harwood, London, pp 55–68

Beauduin M, Gueulette J, De Coster BM, Gregoire V, Octave-Prignot M, Vynckier S, Wambersie A (1989a) Radiobiological intercomparisons of fast neutron beams used in therapy. Strahlenther Onkol 165:263–267

Beauduin M, Gueulette J, Vynckier S, Wambersie A (1989b) Radiobiological intercomparisons of clinical neutron beams for growth inhibition in *Vicia faba* bean roots. Radiat Res 117 (2):245–250

Beauduin M, Laublin G, Octave-Prignot M, Gueulette J, Wambersie A (1992) RBE variation between p(75)+Be and d(50)+Be neutrons determined for chromosome aberrations in *Allium cepa*. Radiat Res 130:275–280

Bewley DK, Cullen BM, Astor M et al (1989) Changes in biological effectiveness of the neutron beam of Clatterbridge (62 MeV p on Be) measured with cells in vitro. Br J Radiol 62:344–347

Field SB, Hornsey S, Kutsutani Y (1986) Effects of fractionated irradiation on mouse lung and a phenomenon of slow repair. Br J Radiol 49:700–707

Gabe M (1968) Techniques histologiques. Masson, Paris, pp 190–191

Geraci JP, Jackson KL, Christensen GM, Thrower PD, Weyer BJ (1977) Mouse testes as a biological test system for intercomparisons of fast neutron therapy beams. Radiat Res 71:377–386

Gregoire V, Beauduin M, Gueulette J, De Coster BM, Octave-Prignot M, Vynckiers S, Wambersie A (1993) Radiobiological intercomparisons of p(45)+Be and p(65)+Be neutron beams for lung tolerance in mice after single and fractionated irradiation. Radiat Res 133:27–32

Gueulette J (1995) Preclinical radiobiological experiments. In: Linz U (ed) Ion beam in tumor therapy. Chapman and Hall, Weinheim, pp 73–82

Gueulette J, Gregoire V, Wambersie A (1994) Courbes de survie cellulaires; facteurs d'influence, modèles et concepts. J Chim Phys 91:109–1106

Gueulette J, Beauduin M, Gregoire V, Vynckier S, De Coster BM, Octave-Prignot M, Wambersie A, Strijkmans K, De Scrijver A, El-Akkad S, Böhm L, Slabbert JP, Jones DTL, Maughan R, Onoda J, Yudelev M, Porter AT, Powers WE, Sabattier R, Breteau N, Courdi A, Brassart N, Chauvel P (1996) RBE variations between fast neutron beams as a function of energy. Intercomparison involving 7 neutron therapy facilities. Bull Cancer Radiother 83 Suppl 1:55 s–63 s

Guichard M, Gueulette J, Meulders JP, Wambersie A, Malaise EP (1980) Biological intercomparison of d(50)+Be and p(75)+Be neutrons. Br J Radiol 53:991–995

Hall EJ, Cavanagh J (1967) The oxygen effect for acute and protracted radiation exposures measured with seedlings of *Vicia faba*. Br J Radiol 40:128–133

Hall EJ, Kellerer A (1979) Review of RBE data for cells in culture. In: Barendsen GW, Broerse J, Breur K (eds) High LET radiations in clinical radiotherapy. Pergamon, Oxford, pp 171–174

Hall EJ, Astor M, Brenner DJ (1992) Biological intercomparisons of neutron beams used for radiotherapy generated by p-Be in hospital-based cyclotrons. Br J Radiol 65:66–71

Hartmann G, Menzel HG, Schuhmacher H (1981) Different approaches to determine effective quality factors and dose equivalent using the Rossi counter. In: Burger G, Ebert HG (eds) Proceedings of the 4th Symposium on Neutron Dosimetry, Munich, EUR7448 CEC, Luxembourg, pp 225–236

International Commission on Radiation Protection (ICRP) (1963) Report of the RBE Subcommittee to the International Comission on Radiation Protection and the International Commission on Radiation Units and Measurements. Health Phys 9:357–386

International Commission on Radiation Units and Measurements (ICRU) (1978) Dose specification for reporting external beam therapy with photons and electrons. ICRU, Bethesda (ICRU report 29)

International Commission on Radiation Units and Measurements (ICRU) (1979) Quantitative concepts and dosimetry in radiobiology. ICRU, Bethesda (ICRU report 30)

International commission on Radiation Units and Measurements (ICRU) (1986) The quality factor in radioprotection. ICRU, Bethesda (ICRU report 40)

Joiner MC (1989) A comparison of the effects of p(62)+Be and d(16)+Be neutrons in mouse kidney. Radiother Oncol 13:211–214

Joiner MC, Field SB (1988) The response of mouse skin to irradiation with neutrons from the 62 MeV cyclotron of Clatterbridge, UK. Radiother Oncol 12:153–166

Joiner MC, Maughan RL, Fowler JF, Denekamp J (1983) The RBE for mouse skin irradiated with 3-MeV neutrons: single and fractionated doses. Radiat Res 95:130–141

Kellerer AM, Rossi HH (1972) The theory of dual radiation action. Curr Top Radiat Res Q 8:85–157

Markus B, Kolliopoulou H (1965) Ein qualitativer biologischer Wirkungsunterschied nach Bestrahlung von Teilungsgewebe mit schnellen Elektronen verschiedenen Energiespektrums. Biophysik 2:263–270

Menzel HG, Pihet P, Wambersie A (1990) Microdosimetric specification of radiation quality in neutron radiation therapy. Int J Radiat Biol 57:865–883

Mijnheer BJ, Battermann JJ, Wambersie A (1987) What degree of accuracy is required and can be achieved in photon and neutron therapy? Radiother Oncol 8:237–252

Pihet P (1989) Etude microdosimétrique de faisceaux de neutrons de haute énergie. Applications dosimétriques et radiobiologiques. Thesis, Catholic University Louvain, Louvain-la-Neuve

Pihet P, Gueulette J, Menzel HG, Grillmaier RE, Wambersie A (1988) Use of microdosimetric data of clinical relevance in neutron therapy planning. Radiat Protect Dosimetry 23:471–474

Pihet P, Menzel HG, Schmidt R, Beauduin M, Wambersie A (1990) Biological weighting function for RBE specification of neutron therapy beams. Intercomparison of 9 European centres. Radiat Protect Dosimetry 31:437–442

Puck TT, Markus PI (1956) Action of x-rays on mammalian cells. J Exp Med 103:653–666

Van Dam J (1984) Radiobiological characteristics of high-LET radiations. Thesis, Catholic University, Leuven

Van Dam J, Wambersie A (1981) OER and RBE variation between p(75)+Be and d(50)+Be neutron beams. Br J Radiol 54:921–922

Varma MN, Bond VP (1982) Empirical evaluation of cell critical volume dose vs cell response function for pink mutations in *Tradescantia*. Proceedings of the 8th Symposium on Microdosimetry, Jülich, EUR 8395 CEC, Luxembourg, p 439–450

Wambersie A (1992) Neutron therapy from radiobiological expectation to clinical reality. Radiat Protect Dosimetry 44:379–395

Wambersie A, Battermann JJ (1987) Practical problems related to RBE in neutrontherapy. In: Kärcher KH, Kogelnik HD, Szepesi T (eds) Progress in radio-oncology III. ICRO (International Club for Radio-Oncology), Vienna, pp 155–162

Wambersie A, Menzel HG (1993) RBE in fast neutron therapy and in boron neutron capture therapy. A useful concept or a misuse? Strahlentherapie 169:57–64

Wambersie A, Dutreix J, Gueulette J, Lellouch J (1978) Early recovery for intestinal stem cells as a function of the dose per fraction, evaluated by survival rate after fractionated irradiation of the abdomen of mice. Radiat Res 58:498–515

Wilson KSJ, Field SB (1970) Further measurement of LET spectra with a 10 cm spherical Rossi counter. In: Ebert HG (ed) Proceedings of the 2nd Symposium on Dosimetry, EUR 4452. CEC, Luxembourg, p 137

Withers HR, Elkind MM (1970) Microcolony survival assay for cells of mouse intestinal mucosa exposed to radiation. Int J Radiat Biol 17:261–267

Zaider M, Brenner DJ (1985) On the microdosimetric definition of quality factors. Radiat Res 103:302–316

Zywietz F, Menzel HG, Van Beuningen D, Schmidt R (1982) A biological and microdosimetric intercomparison of 14 MeV d-T neutrons and 6 MeV cyclotron neutrons. Int J Radiat Biol 42:223–228

Relative Biological Effectiveness of Neutrons for Cancer Induction and Other Late Effects: A Review of Radiobiological Data

H. Engels[1,2] and A. Wambersie[2]

[1] Department of Radiation Protection, Studiecentrum voor Kernergie (SCK/CEN), Boeretang 200, B-2400 Mol, Belgium
[2] Université Catholique de Louvain, Cliniques Universitaires St-Luc, Avenue Hippocrate, 54, B-1200 Brussels, Belgium

Abstract

The risk of secondary cancer induction after a therapeutic irradiation with conventional photon beams is well recognised and documented. However, in general, it is totally overwhelmed by the benefit of the treatment. The same is true to a large extent for the combinations of radiation and drug therapy. After fast neutron therapy, the risk of secondary cancer induction is greater than after photon therapy. This can be expected from the whole set of radiobiological data, accumulated so far, which shows systematically a greater relative biological effectiveness (RBE) for neutrons for all the biological systems which have been investigated. Furthermore, the neutron RBE increases with decreasing dose and there is extensive evidence that neutron RBE is greater for cancer induction and for other late effects relevant in radiation protection than for cell killing at high doses as used in therapy. Almost no reliable human epidemiological data are available so far, and the aim of this work is to derive the best risks estimate for cancer induction after neutron irradiation and in particular fast neutron therapy. Animal data on RBE for tumour induction are analysed. In addition, other biological effects are reviewed, such as life shortening, malignant cell transformation in vitro, chromosome aberrations, genetic effects. These effects can be related, directly or indirectly, to cancer induction to the extent that they express a "genomic" lesion. Since neutron RBE depends on the energy spectrum, the radiation quality has to be carefully specified. Therefore, the microdosimetric spectra are reported each time they are available. Lastly, since heavy-ion beam therapy is being developed at several centres worldwide, the available data on RBE at low doses are reviewed. It can be concluded from this review that the risk of induction of a secondary cancer after fast neutron therapy should not be greater than 10–20 times the risk after photon beam therapy. For heavy ions, and in particular for carbon ions, the risk estimate should be divided by a factor of about 3 due to the reduced integral dose. The risk has to be balanced against the expected improvement in cure rate when the indication for high-LET therapy has been correctly evaluated in well-selected patient groups.

Recent Results in Cancer Research, Vol. 150

Introduction

Accurate knowledge of the neutron relative biological effectiveness (RBE) for cancer induction is of paramount importance in radiation protection for assessing the risk related to a neutron exposure and for setting appropriate dose limits.

In radiation therapy also, it is essential to know, as accurately as possible, the neutron RBE for late effects and in particular for cancer induction.

Induction of a secondary cancer by a radiological and/or medical treatment has always been a great source of concern for the radiation oncologist as well as for the medical oncologist. Radiation-induced cancers have indeed been reported after radiation treatment by many authors, in particular for locally extended cervix carcinomas (Day and Boice 1983; Boice et al. 1988), and also after a combination of radiation and drug therapy (de Vathaire et al. 1989; Tucker et al. 1988; Curtis et al. 1992). However, in most cases, the benefit of the treatment (in terms of survival, local control and quality of life) largely overwhelms the risk of secondary cancer inducation after photon irradiation. As an example, Fig. 1 and Table 1, taken from an International Agency for Research on Cancer (IARC) study, give the risk of cancer associated with a broad range of organ doses, after radiation therapy for cervical cancer. A total number of 150000 patients with cervical cancer were followed, and 4188 secondary cancers were observed. Table 2 illustrates the increased risk after combination of radiation therapy and chemotherapy.

The risk of secondary cancer induction, after fast neutron therapy, is greater than after conventional photon therapy, as can be anticipated from the available radiobiological data indicating high RBE values for cancer inducation after neutron irradiation. However, as in photon therapy, for well-selected groups of patients, the benefit due to the treatment with high linear energy transfer (LET) radiations is definitely higher than the increased risk of secondary cancer induction. It is one of the reasons why the clinical indications for high-LET radiation therapy need to be well evaluated; this problem is discussed at length in the different chapters of the present book.

As a generally practised approach, high-LET radiations should not be applied to children because of the increased risk of secondary cancer induction. This is important to consider if long survival is expected (Wambersie et al. 1982).

In conventional photon therapy, patients are also exposed to low doses of neutrons when the beam energy exceeds about 10 MV (Karzmark 1987; Karzmark et al. 1993).

The production of neutrons from high-energy medical accelerators has been evaluated by Nath et al. (1984). Firstly, neutrons are produced by high-energy photons incident on the various materials of the target, flattening filter, collimation system and other essential components of the machine. For a given photon observed dose to the target volume, the neutron yield increases rapidly with the energy of the photons from 10 to 20 MV, but remains rather

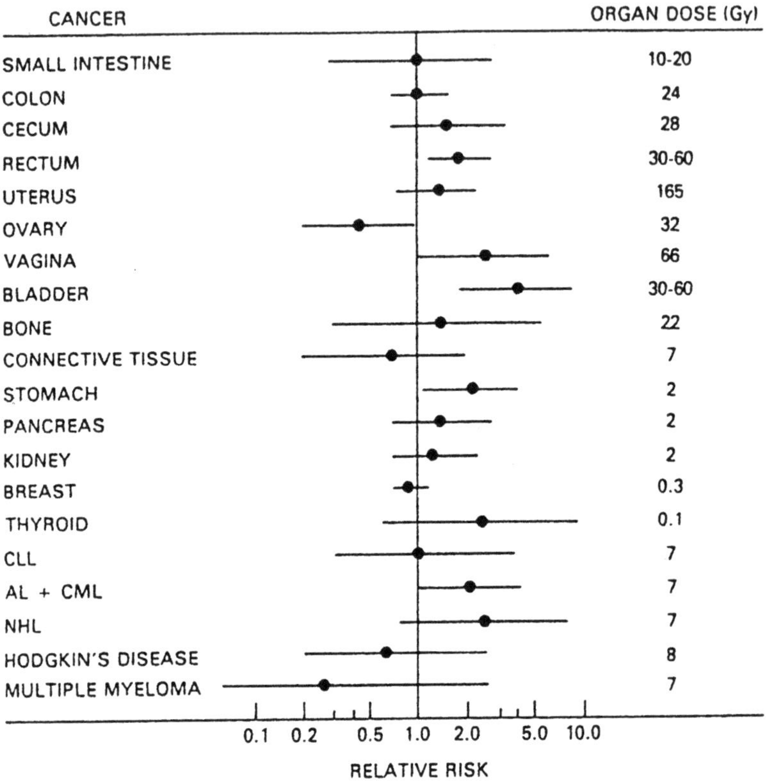

Fig. 1. Results of the IARC study on cervical cancer. Relative riks (RR) of developing a second primary cancer in selected organs associated with radiotherapy for cervical cancer. The 90% confidence intervals and the average organ doses (in gray) are indicated. *CLL*, chronic lymphocytic leukaemia; *AL+CML*, acute leukaemia of all types and chronic myelogenous leukaemia; *NHL*, non-Hodgkin's lymphoma. (From Boice et al. 1988)

constant above this energy (Fig. 2). Secondly, neutrons are produced from high-energy photons within the patient's tissues. These neutrons are distributed in virtual conformance with the photon fluence within the irradiated volume (Fig. 3). In the authors' experience, for typical equipment, the integral doses from accelerator-produced neutrons are about $4\text{–}7\times10^{-3}$ kg×Gy, per gray to the target volume. This corresponds (for a 70-kg patient) to an average body dose of 0.6–1 mGy for a target dose of 10 Gy. The authors conclude that, for radiotherapy patients, the risk of cancer induction by accelerator-produced neutrons is negligible. However, this conclusion has been questioned (e.g. Hall 1996), and for the selection of the optimal photon energy (e.g. between 10 and 20 MV), the benefit of improving the physical selectivity of the irradiation by increasing the photon energy has to be balanced against the corresponding increase of risk of cancer induction due to the high RBE of neutrons for cancer induction (Wambersie and Gahbauer 1996).

Table 1. Organ-specific risk estimates of second primary cancer after radiotherapy for cervical cancer. Summary of the IARC study (Boice et al. 1988)

Second cancer (ICDO)	RR at 1 Gy[a, b]		Attributable risk (%)[c]	
	Risk	90% CI	Risk	90% CI
Colon (153, 159, 0)	1.00	(0.99, 1.02)	2.0	(–43,38)
Caecum (153.4)	1.02	(0.99, 1.09)	35.1	(–43, 71)
Rectum (154)	1.02	(1.00, 1.04)	43.5	(4.8, 67)
All female genital (182–184)	1.01	(1.00, 1.02)	33.3	(–11, 62)
Ovary (183)	1.01	(0.98, 1.14)	28.6	(–100, 82)
Vagina (184.0)	1.03	(1.00, 1.08)	62.3	(0.0, 84)
Other genital (184.1, 184.8)	0.98	(0.95, 1.07)	–22.0	(–100, 44)
Bladder (188)	1.07	(1.02, 1.17)	75.3	(47, 88)
Connective tissue (171)	0.95	(0.89, 1.13)	–49.3	(100, 47)
Stomach (151)	1.69	(1.01, 3.25)	57.8	(2.0, 82)
Pancreas (157)	0.00	(0.72, 1.62)	0.0	(–100, 54.1)
Kidney (189)	1.71	(1.03, 3.24)	58.5	(5.7, 82)
Breast (174)	1.03	(0.13, 2.29)	1.0	(–37, 29)
Thyroid (193)	13.30	(0.00, 77.0)	57.4	(–66, 89)
Leukemia				
CLL[d]	1.00	(0.90, 1.43)	2.9	(–100, 74)
AL+CML[d]	1.149	(1.00, 1.45)	50.5	(0.0, 76)

[a] Risk estimates were computed as the "average" increase in relative risk for each cancer, specific for 10-year survivors and ages at exposure for that cancer, unless otherwise noted. Estimates were not feasible for cancers of the small intestine, uterus, bone, Hodgkin's disease, non-Hodgkin's lymphoma and multiple myeloma because of the small number of these cancers observed in the non-exposed patients.

[b] Computed as the excess RR per Gy+1, i.e. (RR-1)/D+1, where RR is the overall RR for 10-year survivors, unless otherwise noted, and D is the average organ does in gray.

[c] [(RR–1)/RR]. 100% where RR is the overall RR for the appropriate 1+ or 10+ year latency intervall.

[d] Estimates for 1-year survivors only. Estimates for 10-year survivors were not made because either the small number of cases precluded meaningful analyses or a shorter interval was considered more appropriate, e.g. for leukaemia

Almost no reliable human epidemiological data are available to date concerning the neutron RBE for cancer induction. In fact, it will be very difficult to derive valid conclusions from the follow-up of the Japanese population irradiated in Hiroshima because of the considerable dosimetric uncertainties (existence or not of a significant neutron component). In fast neutron therapy, the choice of treatment technique was often the result of negative clinical selection. In addition, the patient series are rather small and the survival often poor. Therefore, limited conclusions can be expected, if any, from these patient series. There have been few accidents in which neutron exposure was known to be sufficiently large to estimate a risk. One of these was a critical accident at Los Alamos National Laboratory (Hempelmann et al. 1980). Two of the seven long-term survivors died from acute myeloid leukaemia 19 and 33 years after exposure.

Table 2. Risk of induction of acute nonlymphocytic leukaemia and myelodysplastic syndrome after radiation therapy and/or administration of alkylating agents. (From Curtis et al. 1992)

Radiation therapy	Alkylating agents	Case patients	Controls	Matched relative risk	95% CI
No	No	20	138	1.0[a]	–
Yes	No	12	51	2.4	1.0–5.8
No	Yes	25	32	10.0	3.9–25.2
Yes	Yes	23	14	17.4	6.4–47.0

[a] The reference group.

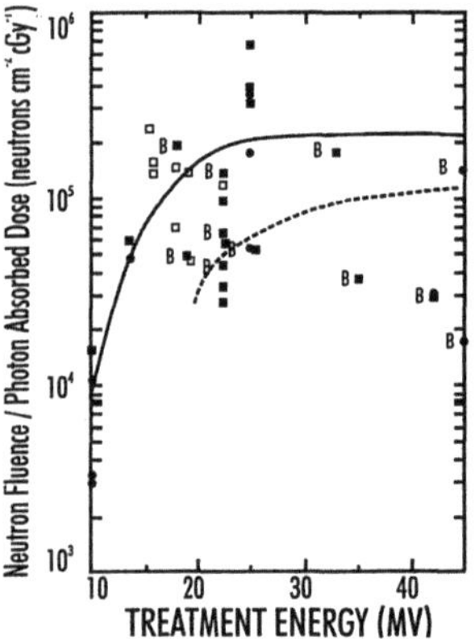

Fig. 2. Neutron fluence as a function of the photon beam energy (for a given dose to the target volume). *Solid curve* corresponds to the case in which all neutron-producing components are made of tungsten (alone or in combination with Pb) and represent the maximum possible neutron fluence if the electron beam strikes only these materials. *Dotted curve*, for comparison, is for all neutron sources being of Cu. *Points* are representative measurements for several accelerator types (*B*, betatrons). (After Nath et al. 1984)

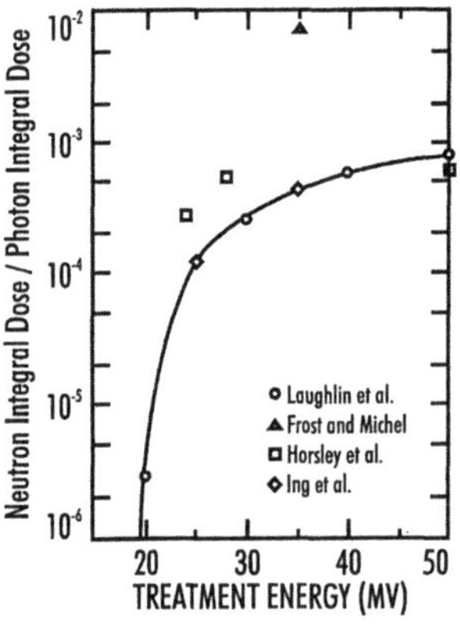

Fig. 3. Estimates of integral dose due to neutrons produced within the patient tissues and expressed as a fraction of the photon integral dose (delivered for the treatment). *Abscissa*, photon beam energy in MV. The curve is drawn through points from Laughlin et al. (1979). Estimates from other authors are indicated for comparison. (After Nath et al. 1984)

As a consequence of the absence of human observation, animal data have to be evaluated very accurately, and as usual extrapolation to the human situation has to be made with the greatest care.

In addition to tumour induction, other biological effects are reviewed in this chaper: e.g. life shortening, malignant cell transformation in vitro, chromosome aberrations, genetic effects. Some of these effects can be related, directly or indirectly, to cancer induction (they can be considered as expressing a "genomic" lesion). These data obtained for different systems and effects contribute to a broader and safer knowledge of the neutron RBE at low dose, and therefore contribute to support, and to some extent validate, the neutron RBEs obtained for tumour induction.

The neutron RBE depends on the biological system, on the dose level but also on the radiation quality (and of course on the irradiation conditions). In the case of neutrons, the radiation quality can be expressed in terms of the neutron energy spectrum but, as a promising alternative, another approach is to provide their microdosimetric characteristics (Wambersie and Menzel 1997). Indeed, systematic microdosimetric studies were performed for both radiation therapy purposes and for protection purposes. In this chapter, for each type of biological study, the microdosimetric spectra of the radiation beams will be given when available.

Since heavy-ion beam therapy (high LET) is being developed at several centres worldwide, the available RBE data, at low doses, are reviewed later in this chapter.

Factors Influencing the Reported Neutron RBE Values

RBE and Biological Criterion

When evaluating the RBE of neutrons for radiotherapy applications, the main biological effect to take into account is cell killing and the consequences of cell killing. Indeed a large amount of data has been accumulated concerning the RBE of fast neutrons for endpoints relevant to fast neutron therapy, i.e. cell survival for in vitro and in vivo systems, early and late tolerance of normal tissues (in addition to effects on tumour cells).

In radiation protection, a larger range of effects has to be considered. For exposure at large doses (accidental, occupational, military exposures), acute/early effects are observed which are to some extent similar to the effects encountered in neutron therapy, and similar RBE values can be derived. In contrast, at lower doses, the most important effects to fear are cancer induction and genetic effects.

In general, the neutron RBEs are lower for the effects relevant in radiation therapy, which are the (direct or indirect) consequences of cell killing. The neutron RBEs are significantly higher for the effects produced at low doses which are relevant in protection (mainly cancer induction).

RBE and Dose Level: Mathematical Models

For the effects relevant in radiation protection (tumour induction, chromosome aberrations, mutations, etc.), the available experimental data are compatible with a dose-effect relation of the linear-quadratic type, between the effect E and the absorbed dose D, for low-LET radiations:

$$E = \alpha D + \beta D^2 \tag{1}$$

The relative importance of the linear and quadratic terms depends on the radiation quality, and for high-LET radiations the equation is practically reduced to:

$$E = \alpha D \tag{2}$$

The linear-quadratic model is also the most widely used model in radiation therapy to express the relation between cell survival and radiation dose $S = e^{-(\alpha D + \beta D^2)}$. In particular it is proven to be useful to predict the effect of dose fractionation in photon therapy.

When decreasing dose, the RBE of neutrons (high LET) relative to gamma rays (low LET) increases, mainly because of the term βD^2 (Eq. 1), which decreases with decreasing dose.

At very low doses, the term βD^2 tends to zero and the RBE tends towards the ratio of the α-coefficients for neutrons α_n and for gamma-rays α_γ. The RBE thus reaches a plateau value when the dose-effect curves, for both radiation qualities, become indistinguishable from their tangents at the origin. This maximum plateau value of the RBE will be designated RBE_{max}:

$$RBE_{max} = \alpha_n / \alpha_\gamma$$

RBE_{max} is an important factor in radiation protection.

In a limited dose range, and in a first approximation, one can assume, for the neutrons, $E = \alpha D_n$ and, for gamma-rays, $E = \beta D_\gamma^2$. In these conditions, the neutron RBE is proportional to $\sqrt{\frac{1}{D_n}}$ and the relation between RBE and neutron dose can be represented, in logarithmic coordinates, by a straight line with a slope –1/2.

This assumption has been confirmed by a review of the available experimental data (Kellerer and Rossi 1972), which found a straight-line relationship, in logarithmic coordinates, between RBE and neutron dose with a slope close to –1/2 (at least for some of the results for part of the investigated dose range) (Fig. 4).

Radiation Quality: Neutron Energy Spectrum and Microdosimetry

Besides the biological criterion and the dose level, neutron RBE also depends on radiation quality, i.e. neutron energy spectrum and reference radiation quality.

The neutron energy, in terms of spectrum or average value, is primary information. Neutrons deposit energy through secondary charged particles

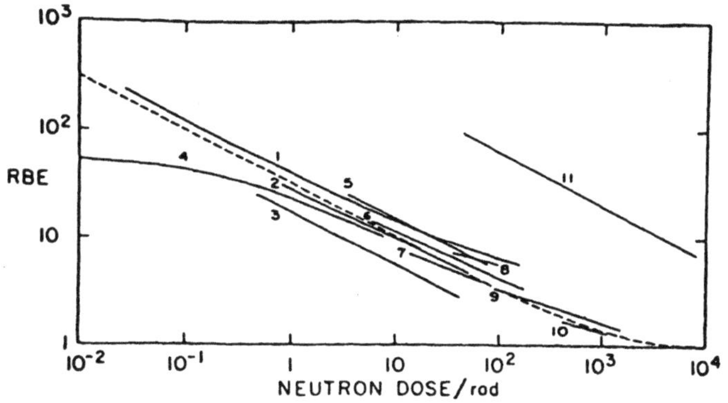

Fig. 4. RBE as a function of neutron dose for various biological effects and different neutron energies. The biological effects (11 effects), the neutron energies and the corresponding references are indicated below. The *dashed curve* has been derived from the Kellerer and Rossi theory. (After Kellerer and Rossi 1972).

Batemann et al. (1972)	*1,* opacification of the murine lens (430 keV) *2,* opacification of the murine lens (1.8 MeV) *3,* opacification of the murine lens (14 MeV)
Sparrow et al. (1972)	*4,* mutations of *Tradescantia* stamen hairs (blue to pink) (430 keV)
Vogel (1969)	*5,* mammary neoplasm in the Sprague-Dawley rat (fission neutrons)
Biola et al. (1971)	*6,* chromosome aberrations in human lymphocytes (fission neutrons)
Hall (1972)	*7,* growth reduction of *Vica faba* root, aerated (3.7 MeV) *8,* growth reduction of *Vica faba* root, anoxic (3.7 MeV)
Field (1969)	*9,* skin damage (human, rat, mouse, pig) (6 MeV)
Withers et al. (1970)	*10,* inactivation of intestinal crypt cells in the mouse (14 MeV)
Smith et al. (1968)	*11,* various effects on seeds of *Zea mays* (fission neutrons)

whose distributions critically depend on the shape of the neutron spectrum and on the types and thresholds of nuclear reactions with the different tissues. The biological properties of neutron beams are therefore more directly correlated with the secondary radiation spectra and only indirectly with the neutron energy. However, the spectra of secondary charged particles are not easily accessible information, especially for the fast neutron beams used in therapy.

The measurement of microdosimetric spectra offers an accurate description of the secondary radiation components and provides comprehensive and detailed information on radiation quality. This was proven particularly useful in the field of fast neutron therapy, where a close correlation was derived between the microdosimetric characteristics of the beams and their RBEs for some well-defined biological endpoints.

Fig. 5 compares the microdosimetric spectra [$y{\cdot}d(y)$] obtained for therapeutic neutron beams produced by bombarding a beryllium target with 14 MeV deuterons [d(14)+Be] and 65 MeV protons [p(65)+Be], i.e. the lowest and highest beam energies used in fast neutron therapy, respectively (Tubiana et al. 1990). On the other hand, Fig. 6 compares $y{\cdot}d(y)$ vs. y spectra for neutron energies from 0.55 to 15 MeV (Rodgers and Gross 1974).

A systematic evaluation of the RBE variation with neutron energy, in the range of 0.1–1 MeV, has been performed by Hall et al. (1975) for two biologi-

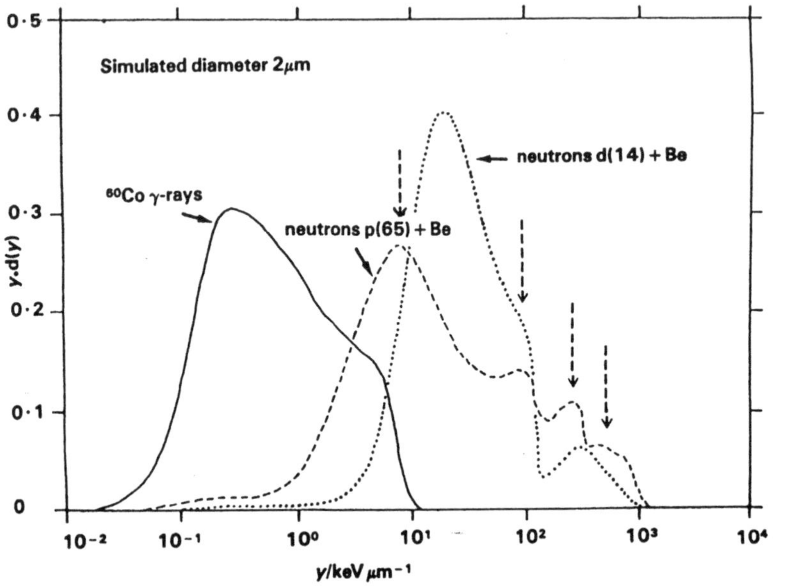

Fig. 5. Comparison of the energy-deposition event spectra for therapeutic neutron beams produced by 14-MeV deuterons and 65-MeV protons on a beryllium target (i.e. the lowest and the highest beam energy used in neutron therapy, respectively). The spectrum obtained for ^{60}Co is shown for comparison. The curves indicate distributions of individual energy-deposition events in a simulated sphere of tissue 2 μm in diameter; the parameter y (lineal energy) represents the energy deposited by a single charged particle traversing the sphere, divided by the mean cord length. The maximum with gamma-rays is at 0.3 keV/μm and with d(14)+Be neutrons at 20 keV/μm. The spectrum for p(65)+Be neutrons shows four peaks (*vertical arrows*): the first is at 8 keV/μm and corresponds to high-energy protons; the second at 100 keV/μm corresponds to low-energy protons; the third at 300 keV/μm is due to alpha-particles; and the last one at 700 keV/μm due to recoil nuclei. (From Tubiana et al. 1990)

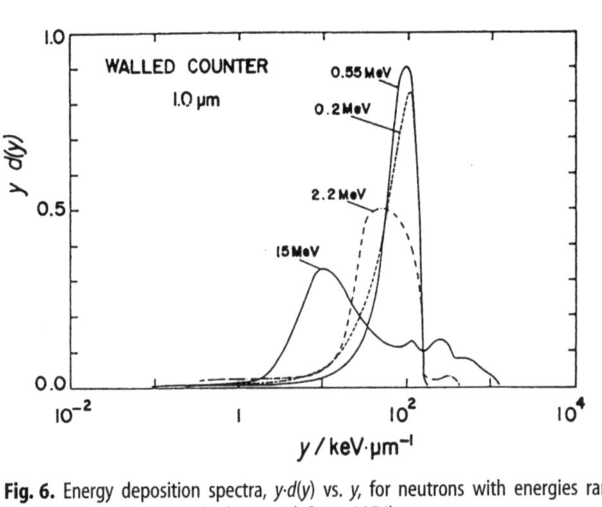

Fig. 6. Energy deposition spectra, $y{\cdot}d(y)$ vs. y, for neutrons with energies ranging from 0.55 to 15 MeV. Site diameter is 1 μm. (From Rodgers and Gross 1974)

cal endpoints: growth delay in *Vicia faba* and lethality in Chinese hamster cells (Fig. 7). In that range, the RBE variation is relatively small; the RBE passes through a maximum value for neutrons of 0.3–0.5 MeV and it will therefore be slightly lower for fission neutrons.

For fast neutrons, in the energy range used for therapy applications, a large amount of data on RBE has been accumulated. As mentioned earlier, they were obtained mainly for mammalian cell lethality and for early and late tolerance of normal tissues. As an example, for 15 MeV neutrons produced by a (d+T) generator, van der Kogel (1979) observed RBE values, relative to gamma-rays, increasing from 1.5 to 9 when dose decreases from 10 to 1 Gy and depending on the biological system and endpoint. Some of his data are presented in Fig. 8.

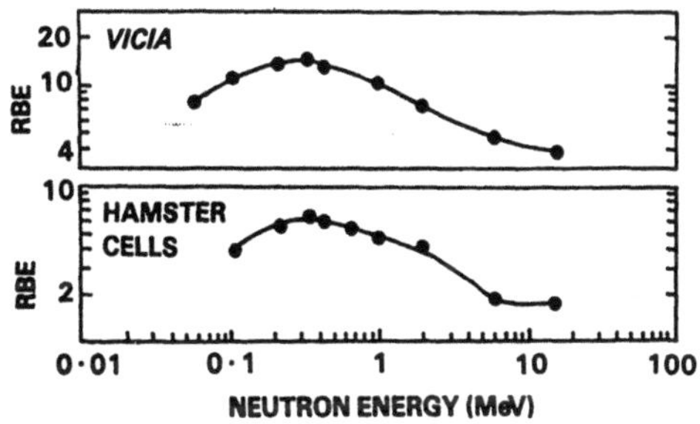

Fig. 7. RBEs of neutrons as a function of energy. Two biological systems are compared: (a) inhibition of growth in *Vicia faba* and (b) lethality of Chinese hamster cells. For both systems, the RBE shows a maximum at 0.3–0.5 MeV, but there is little variation between 0.1 and 1 MeV. (From Hall et al. 1975)

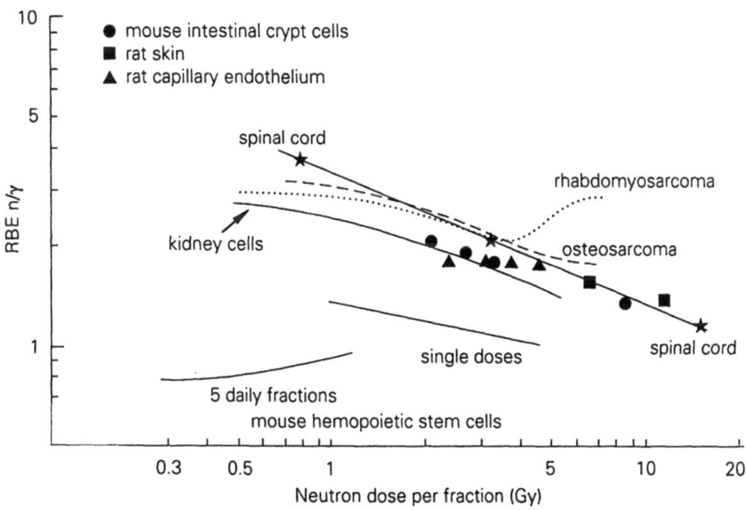

Fig. 8. RBEs of 15-MeV neutrons [from a (d+T) generator], relative to gamma-rays, plotted as a function of neutron dose per fraction for different biological systems in normal tissues and tumours. As a general rule, the neutron RBE increases when decreasing dose per fraction. For example, for late tolerance on spinal cord, the RBE increases from 1.2 to 3.7 when neutron dose per fraction decreases from 16 to 0.8 Gy. (Redrawn from van der Kogel 1979)

In addition to RBE determinations performed in the different neutron therapy centres, direct radiobiological intercomparisons were performed; the largest one was initiated under the auspices of the European Hadron Therapy Group (Breteau et al. 1996). Using intestinal crypt regeneration as the biological system, Gueulette et al. (1996) found a decrease in RBE greater than 20% between the lowest and the highest energy beams used in neutron therapy [i.e. d(16)+Be and p(65)+Be, respectively] (Fig. 9, Table 3). The RBE of p(65)+Be neutrons relative to cobalt-60 gamma-rays was 3.1 for intestinal crypt regeneration in mice.

A general relation between RBE and LET was derived by Barendsen (1968, 1996) for survival of mammalian cells of human origin at different survival (or dose) levels (Fig. 10). The RBE decrease when LET increases beyond about 100 keV/μm is interpreted as being due to the "overkill effect". This ef-

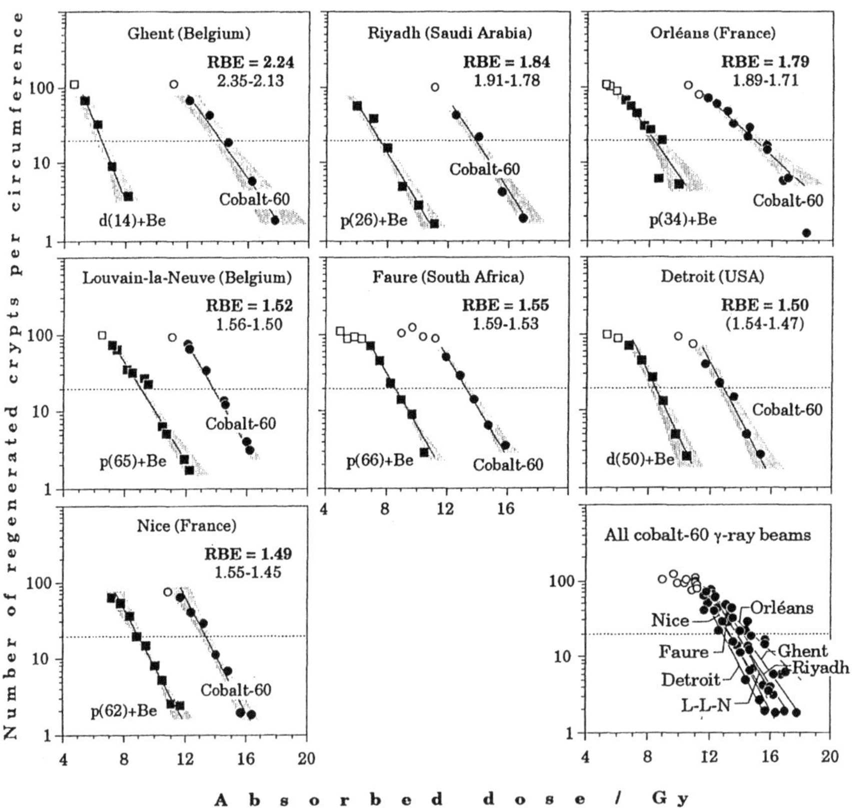

Fig. 9. Radiobiological intercomparisons between seven neutron therapy beams, using intestinal crypt regeneration in mice as the biological system. The dose-effect relationships, after neutron and cobalt irradiation, obtained in the different centres are shown, from which the RBE values are derived. In order to illustrate the reliability of the system, the lower rigth-hand panel compares the data obtained with cobalt in the different centres. (From Gueulette et al. 1996)

Table 3. Results of the direct radiobiological intercomparisons performed at seven neutron therapy facilities. (From Gueulette et al. 1996)

Neutron therapy facility and nuclear reaction	No. of fractions	RBE relative to cobalt-60	RBE relative to the p(65)+Be neutron beam of LLN
Orléans p(34)+Be	1	1.79±0.10[b] (depth=5 cm)	1.18±0.09
Riyadh p(26)+Be	1	1.84±0.07 (depth=5 cm)	1.21±0.06
Ghent	1	2.24±0.11 (depth=5 cm)	1.47±0.12
d(14)+Be	2[a]	2.59±0.14 (depth=5 cm)	–
Faure	1	1.55±0.04 (depth=5 cm)	1.02±0.04
p(65)+Be	1	1.50±0.04 (depth=13 cm)	–
Detroit	1	1.50±0.03 (depth=2 cm)	0.99±0.03
d(48)+Be	1	1.51±0.03 (depth=10 cm)	–
Nice	1	1.49±0.05 (depth=2 cm)	0.98±0.04
p(62)+Be	1	1.45±0.05 (depth=7 cm)	–
Louvain-la-Neuve			
p(65)+Be	1	1.52±0.04 (depth=5 cm)	1.00 (ref.)
p(34)+Be	1	– (depth=5 cm)	1.13±0.06
p(45)+Be	1	– (depth=5 cm)	1.05±0.07
p(75)+Be	1	– (depth=5 cm)	0.96±0.05
d(50)+Be	1	– (depth=5 cm)	1.00±0.03

[a] Two fractions separated by a time interval of 3.5 h.
[b] Confidence interval, $p=0.05$.

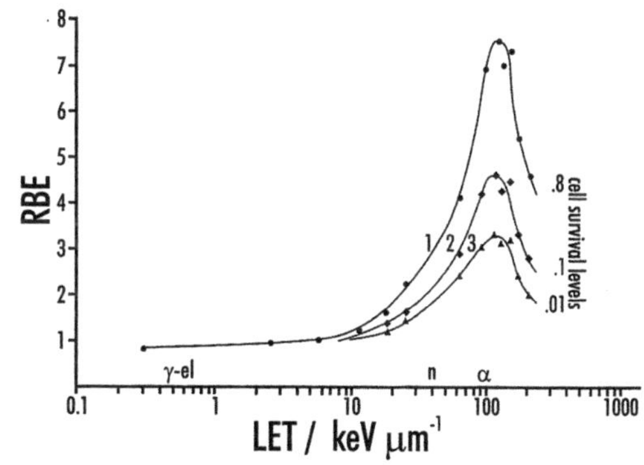

Fig. 10. Variation of RBE with LET for survival of mammalian cells of human origin. The RBE rises to a maximum at an LET of about 100 keV/µm and subsequently falls for higher values of LET. Curves 1, 2 and 3 refer to cell survival levels of 0.8, 0.1 and 0.01, respectively, illustrating that the absolute value of the RBE is not unique but depends on the level of biological damage and thus on the dose level. Typical LET values of gamma-rays, electrons, neutrons and alpha-particles are shown *on the abscissa*. (After Barendsen 1968)

fect should be taken into account when heavy ions are involved. An interesting point made by Barendsen (1979) is that the variation of RBE with LET is steeper for chromosome aberrations than for mammalian cell killing. This observation is certainly relevant when trying to assess the most reliable neutron RBE value for cancer induction.

Reference Radiation Quality

RBE values are currently expressed relative to cobalt-60 gamma-rays or to 200- to 250-kV X-rays. In radiation protection, it has been customary to group together all the "low-LET" radiations (X-rays, gamma-rays, electrons, etc.) and to give them the same value (unity) of the weighting factor W_R (previously the quality factor, Q; ICRP 1991).

The RBE of 200-kV X-rays, relative to cobalt-60 gamma-rays, is close to unity at high doses and high dose rates. In current radiotherapy practice, a value of 1.18 (=1/0.85) is usually assumed (ICRU 1978).

However, at low doses and for the effects relevant in radiation protection, much greater differences in RBE must be taken into account. For doses below 1 Gy, Underbrink et al. (1976) found an RBE equal to 2 for the induction of pink mutations in *Tradescantia*. Edwards et al. (1982) obtained an RBE of 3 for the production of dicentric chromosomes in human lymphocytes. For the same criterion, a lower value (RBE=1.5) was found by Fabry et al. (1985).

Finally, the microdosimetric data indicate that an RBE of at least 2 is to be expected for 200-kV X-rays relative to cobalt-60 gamma-rays (ICRU 1983). Figure 11 compares the microdosimetric spectra $y{\cdot}d(y)$ for mono-energetic photon beams with energy ranging from 1250 MeV down to 11.9 MeV (Kliauga and Dvorak 1978). One can observe in the figure that as the photon energy varies, the systematic change in shape of the microdosimetric spectra is disrupted in the energy region of 60–140 keV. This is directly attributable to a transition from the photoelectric effect to Compton scattering as the predominant mode of interaction with the medium.

In conclusion, when reporting an experimental neutron RBE result, and in addition to the reference radiation quality, it is also necessary to specify the

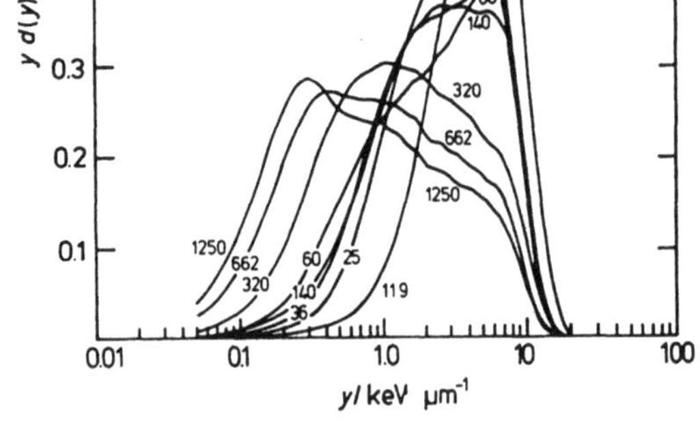

Fig. 11. Microdosimetric spectra $y{\cdot}d(y)$ for monoenergetic photons with various energy (spherical counter of 1 μm simulated diameter). The three lowest energies are produced with a fluorescent X-ray generator. The remainder are emissions from radionuclides: ^{241}Am (59.6 keV, ^{99m}Tc (140 keV), ^{51}Cr (323 keV), ^{137}Cs (662 keV) and ^{60}Co (1170 and 1330 keV). (From Kliauga and Dvorak 1978)

dose rate and fractionation applied. Indeed, it is well known that the effectiveness of a low-LET irradiation decreases with fractionation and dose rate. The reduction in effectiveness can range between 2 and 10 depending on the conditions and the biological endpoint (UNSCEAR 1988). As an example, Grahn and Fritz (1984) have studied the influence of dose rate and fractionation on life shortening in the B6CF1 mouse. For irradiation with gamma-rays, the life shortening at high dose rate was 43 days/Gy and was reduced to 4 days/Gy after exposure at low dose rate. After fractionated irradiation (60 fractions), the life shortening ranged between 11 and 23 days/Gy.

In radiation protection, we are usually concerned with low doses received in many fractions and/or at low dose rate. Obviously, the most relevant RBE values are those obtained in such conditions. However, in practice among the published RBE values, some have been obtained (1) with either X-rays or gamma-rays as the reference radiation quality, (2) delivered at either high or low dose levels and/or (3) at either high or low dose rate. The conditions in which the reference irradiation has been delivered can introduce significant differences in the observed neutron RBE values and can be the cause of much confusion.

Survey of the Experimental Data on Neutron RBE for Tumour Induction

RBE of Fission Neutrons for Tumour Induction

Most of the experimental data were obtained with fission neutrons for obvious reasons of safety of nuclear workers and military applications (Hollander et al. 1982). The RBE of fission neutrons for tumour induction has been studied for several tumour types. Shellabarger et al. (1982) have studied the induction of mammary tumours in the Sprague-Dawley and ACI rat after 0.43 MeV neutron irradiation (Fig. 12). The biological endpoints were the number of radioinduced tumours and their delay in appearance.

The neutron RBE, relative to X- and gamma-rays, was found to be proportional to the inverse square root of the neutron dose $(1/\sqrt{D_n})$ in Sprague-Dawley rats and in ACI rats treated with DES (diethylstilbestrol). At 10 mGy,

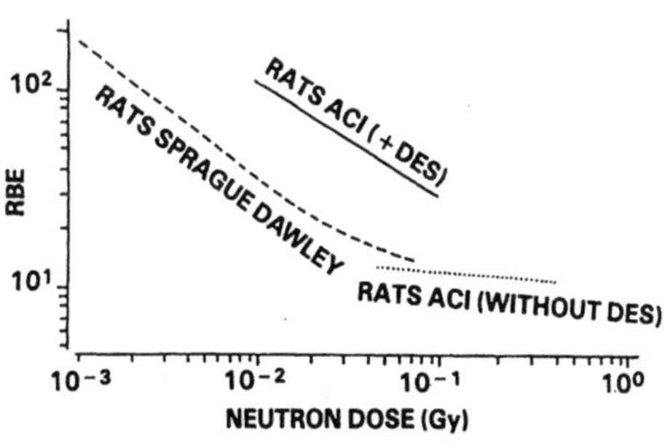

Fig. 12. RBEs of 0.43-MeV neutrons, as a function of neutron dose, for induction of mammary tumours in the rat. For Sprague-Dawley rats, as well as ACI rats after stimulation with oestrogen (diethylstilboestrol, DES), the RBE varies inversely with $\sqrt{D_n}$ over the range of doses explored. At 10 mGy the RBE values are about 50 and 100, respectively. The RBE is lower (about 10) for ACI rats without hormone stimulation. (After Shellabarger et al. 1982)

the RBE reached 50 for induction of fibroadenoma in Sprague-Dawley rats and exceeded 100 for induction of adenocarcinoma in ACI rats treated with DES. For ACI rats, without hormonal stimulation, the RBE remained lower, i.e. about 10.

Lower RBE values were obtained at the Radiobiological Institute TNO in Rijswijk. In WAG/Rij rats, an RBE of 15 was obtained for induction of adenocarcinoma and an RBE of 13 for fibroadenoma. An RBE of 7 was obtained for the induction of fibroadenoma in the Sprague-Dawley rats (Broerse et al. 1985).

Grahn et al. (1983) compared fission neutrons with gamma-rays for induction of different types of tumours and obtained RBE values ranging from 8 to 50 depending on the type of tumour; in particular the RBEs for induction of squamous cell carcinomas were greater than for sarcomas.

For induction of lung and mammary tumours in mice, Ullrich (1983) found RBE values of 18.5 and 33, respectively, at a neutron dose of 25 cGy. Using another method of analysis and assuming for the RBE the inverse square root dependence for induction of lung adenocarcinoma, the same authors inferred an RBE of 71 at a neutron dose of 10 mGy.

As fas as the age at the time of exposure is concerned, the RBE of d(6.3)+Be neutrons (modal energy 3.1 MeV) for tumour induction in young mice has been studied by Maisin et al. (1996). The authors compared the effects of irradiation at 7 or 21 days (i.e. during lactation and at the end of the lactation period, respectively). The RBE for induction of all malignant tumours and for induction of hepatocellular carcinoma (the most frequently induced tumour in these experimental conditions, in C57BL mice) ranged between 5 and 8 (relative to 250-kVp X-rays) (Fig. 13). A lower RBE was observed for leukaemia induction. For life shortening an RBE of 3 was observed.

Again in infant mice, the effects of d(6.3)+Be neutrons on the induction of foci, adenomas and carcinomas in the livers of C57BL mice were studied (Maisin et al. 1995), and in particular the possible synergy between neutrons

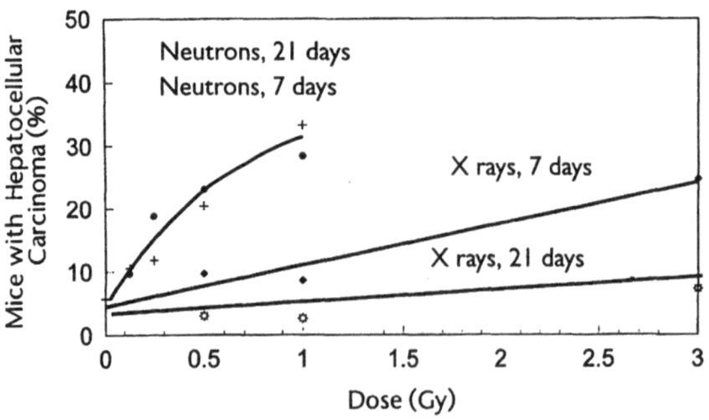

Fig. 13. Induction of hepatocellular carcinoma after neutron irradiation (modal energy: 3.1 MeV; dose range from 0.125 to 1 Gy) and 250 kVp irradiation (dose range from 0.5 to 3 Gy). Young mice, 7 and 21 days old, were irradiated. (From Maisin et al. 1996)

Table 4. RBE of neutrons for tumour induction (neutron energy 0.43–1 MeV; dose level 10 mGy). (After Broerse and van Bekkum 1989)

Tumour	RBE
Lung adenocarcinoma in mice α-neutrons/α-photons	18.5
$D_n^{-½}$	71
Malignant transformation in hamster embryo cells	10–25
Myeloid leukaemia in mice, daily chronic irradiation	16
Myeloid leukaemia in mice, acute irradiation, square root of α-neutrons/α-photons	13
Myeloid leukaemia in mice	3
Fibroadenoma in Sprague-Dawley rats	
$D_n^{-½}$	50
Adenocarcinoma in ACI rats treated with DES,	
$D_n^{-½}$	100
Adenocarcinoma in WAG/Rij rats	15
Fibroadenoma in WAG/Rij rats	13
Fibroadenoma in Spague-Dawley rats	7

and a potent initiator, diethylnitrosamine (DEN). The rate of appearance of foci increased significantly when a dose of 0.125 Gy neutrons was administered 7 days before or after a dose of 1.25 µg DEN. No significant difference was observed depending on whether neutrons were given 7 days prior to or after the DEN. On the other hand, no effect was observed after a combination of DEN and X-irradiation (up to 3 Gy).

Other types of tumours have been studied and large variations in RBE were found depending on the type of tumours: the RBE for fission neutrons ranged from 2–3 to more than 200, but most of the observed RBE values were below 100. The data have been reviewed by Broerse et al. (1989) and are summarized in Table 4.

RBE of Fast Neutrons for Tumour Induction

Studies of the RBE of fast neutrons for tumour induction are particularly relevant in fast neutron therapy to evaluate the risks of late effects. However, few data are available so far, since most of the experiments have used endpoints directly relevant for therapy (see in particular Fig. 8).

Tumour induction and life shortening after single and fractionated neutron and Cs-137 gamma irradiation were studied in C57Bl mice by Maisin et al. (1988). For neutrons produced by bombarding a beryllium target with 50-MeV deuterons (as used for therapy at Louvain-la-Neuve), an RBE between 1 and 1.5 was derived after single fraction irradiation (gamma dose 2–6 Gy) and an RBE of 2 and 3 after fractionated neutron irradiation (eight neutron fractions, 3 h apart). The neutron RBE for life shortening was not significantly different from unity.

The preliminary results obtained at NIRS in Japan by Furuse et al. (1997) indicate an RBE of 33 for d(30)+Be neutrons (mean energy 13 MeV; average LET 10.7 keV/μm) for tumour induction in C57BL/6J mice. The liver tumours were the most frequently induced.

Fractionation or Dose Rate Effect for Neutrons and High-LET Radiations

An important problem to consider is the effect of fractionation and of dose rate on the effectiveness of neutrons for tumour induction. As fas as low-LET radiations are concerned, most of the available data indicate that fractionation and low dose rate reduce the risk of tumour induction. There are few exceptions, such as the data of Maisin et al. (1988) mentioned above, on induction of leukaemia, carcinomas and sarcomas.

In contrast, for high-LET radiations, the effects of fractionation and dose rate depend on the type of tumour (Table 5). For the induction of ovarian tumours in mice, Ullrich (1984) observed that irradiation with fission neutrons was less effective when delivered at low dose rate than at high dose rate. However, for induction of mammary carcinoma, the effect of fission neutrons was enhanced at low dose rates. This was also observed by Vogel and Dickson (1982) for mammary tumours.

Studies performed in Rijswijk on mammary carcinoma in WAG/Rij rats indicated that, for equal total absorbed dose of X-rays and 0.5 neutrons, the tumours *appeared at approximately the same age* after single and fractionated irradiation. As pointed out by Broerse et al. (1985), experimental studies on mammary carcinogenesis are generally based on whole body irradiation of the animals. The induction of mammary cancer can easily be modified by

Table 5. Effect of fractionation or dose rate on tumour induction or life shortening after high-LET irradiation. (After Broerse et al. 1989)

	Change in effectiveness[a]
Mammary tumours	
Ullrich 1984	+
Vogel and Dickson 1982	+
Broerse et al. 1985	=
Ovarian tumours	
Ullrich 1984	–
Pulmonary tumours	
Ullrich 1984	=
Little et al. 1985	+ or –
Lundgren et al. 1987	+
Life shortening in mice	
Thomson et al. 1985	+
Maisin et al. 1988	=

[a] Enhanced (+), reduced (–) or equal (=) with respect to single-dose or high-dose rate.

hormonal factors, and it might well be that specific endocrinological effects caused by the irradiation influence mammary tumour induction to a lesser extent with fractionated or low dose rate exposures.

Life Shortening After Neutron Irradiation

Life shortening in mice, as in mammals in general, after exposure to low doses of radiations is a result exclusively of tumour induction. Therefore, this criterion is particularly useful for evaluating tumour induction as a whole, complementing the study of the induction of the different individual tumour types. Some results on neutron RBE for life shortening were mentioned in the previous section, since they have often been reported, by the authors, together with the results for tumour induction.

A comprehensive review of the available data was performed by the United Nations Scientific Committee on the Effects of Atomic Radiations (UNSCEAR 1982). For X- and gamma-rays, the data were compatible with a linear *and* a linear-quadratic dose-effect relationship (Fig. 14). For neutrons, the dose-effect relationship was concave downwards (Fig. 15). As a result, the RBE depends on dose (or level of effect); in addition it depends on dose rate and fractionation mainly because of the variation of the efficiency of X-rays with the time-dose pattern. From the comparison of the data of Fig. 14 and 15,

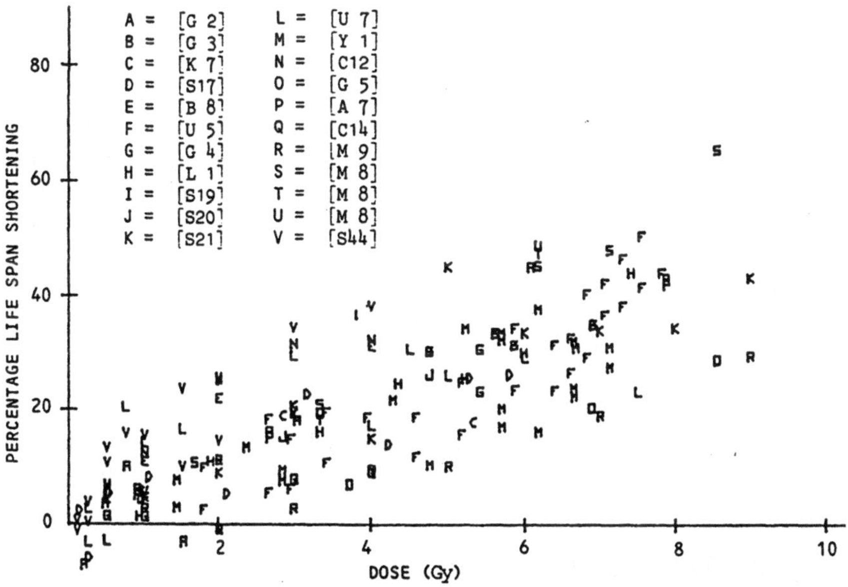

Fig. 14. Dose effect relationship for life shortening in the male and female mouse following single acute exposure to X- and gamma-rays. Review of the available data be UNSCEAR (1982). A total of 35 experimental series performed on 20 strains were analysed (dose range 0–9 Gy, maximum life-span-shortening effect up to 60% of normal). (From UNSCEAR 1982)

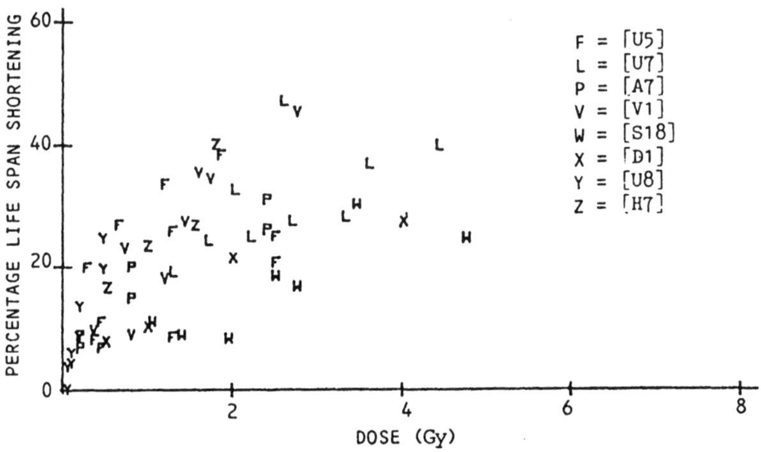

Fig. 15. Dose effect relationship for life shortening in the male and female mouse following single acute exposure to monoenergetic, fission and weapon neutrons. Review of the available data by UNSCEAR (1982). About 14 experimental series performed on 4 strains were analysed (dose range 0–5 Gy, maximum life-span-shortening effect up to 50% of normal). (From UNSCEAR 1982)

following values of the neutron RBE can be deduced for life shortening. For acute doses in the region of a few tenths to a few grays of gamma-rays (and equivalent doses of neutrons), RBE values of 2 to 3 have generally been found. The RBE increases up to 5–6 for short fractionation or when decreasing dose rate down to hundredths of gray of neutrons per day. More consistent changes have been found for extremely long courses of irradiation where RBE values of 9–15 are quoted.

The following average RBE values applicable to exposures for duration of life may be derived: RBE values between 10 and 20 depending on the level of dose or effect. An RBE value of 10 can be expected at 50% life shortening. At dose rates corresponding to 0.01 Gy/day (of low-LET radiation), the neutron RBE would increase up to 40 (UNSCEAR 1982).

Storer and Mitchell (1984) compared a single dose of fission neutrons with gamma-rays at 0.08 Gy/day. The RBE for life shortening increased with decreasing dose and reached, at low dose, a plateau value of 13–22, which may be considered as the RBE_{max}.

For fast neutrons, lower RBE values were found for life shortening. In C57BL/Cnb mice, Maisin et al. (1988) observed an RBE close to unity, for single and fractionated irradiation (eight fractions) with d(50)+Be neutrons.

Malignant Transformation In Vitro

Neutron RBE for the transformation of mammalian cells in vitro has been studied by several teams. For example, Hill et al. (1985) studied the influence of fractionation and dose rate.

The RBE of fission neutrons compared to gamma-rays is relatively low and close to 3 for a single dose. However, fractionation and reduction in dose rate considerably increases the effectiveness of the neutron irradiations (range of doses 0.01–0.1 Gy). The opposite was found with gamma-rays. Consequently, high neutron RBE values were found for some time-dose patterns: $RBE_{max}=35$ if neutrons and gamma-rays were delivered at low dose rates and $RBE_{max}=70$ for fractionated irradiation.

The same conclusions can thus be derived as for the other biological effects discussed above concerning the influence of the time-dose pattern on the neutron RBE.

Chromosome Aberrations

Many studies have been performed on the RBE of different radiation qualities for the production of chromosome aberrations. The criterion most widely used is the number of dicentric chromosomes induced in human lymphocytes; use of this criterion was indeed proven to provide useful information for radiation protection (Geard and Brenner 1997).

More recently, the premature chromosome condensation (PCC) assay has been developed, and is envisaged as an important cytogenetic tool:

- In radiation protection, for biological dosimetry after whole-body and partial body accidental irradiation
- In radiation therapy, in the prediction of human tumour radiosensitivity

Induction of Dicentric Chromosomes

The results of Edwards et al. (1982) are summarized in Table 6. It is assumed that the biological effect (here the number of dicentric chromosomes per cell) is given by the equation: $E=\alpha D+\beta D^2$, D being the absorbed dose. The maximum RBE value, at low dose (RBE_{max}), is then given by the ratio of the α-parameters for the two radiation qualities. The results obtained by Fabry et al. (1985) are presented for comparison in Table 7.

The results can be summarized as follows. For 250-kV X-rays, relative to Co-60 gamma-rays, Edwards (Edwards et al. 1982) found an RBE_{max} of 3.03 and Fabry one of 1.5. Edwards also found very low RBE values for 15-MeV electrons (0.35–0.57).

For fission neutrons (mean energy 0.7–0.9 MeV), Edwards obtained an RBE_{max} of about 50. Bauchinger et al. (1984) reported a lower RBE value for fission neutrons, RBE = 11 at 0.04 Gy. However, in his experiments, the RBE continued to rise when decreasing dose for the lowest dose values which were studied, which implies a higher RBE_{max} value.

For the fast neutrons used for therapeutic applications, the RBE is significantly lower than for fission neutrons. Edwards reported an RBE value of 16.7 for 14.7-MeV neutrons and 30.3 for cyclotron-produced neutrons with a

Table 6. Chromosome aberrations in human lymphocytes for different radiation qualities. (From Edwards et al. 1982)

Radiation	α (Gy^{-1})×10	β (Gy^{-2})×10^2	RBE$_M$ [a]
15-Mev electrons	0.55±1.12	5.66±0.55	0.35
15-MeV electron (pulsed)	0.90±1.92	6.08±0.92	0.57
^{60}Co γ-rays	1.57±0.29	5.00±0.20	1.00
250-kV X-rays	4.76±0.54	6.19±0.31	3.03
14.7-MeV neutrons	26.2±4.0	8.8±2.8	16.7
Cyclotron neutrons	47.8±3.3	6.4±2.0	30.3
^{252}Cf neutrons (2.13 Mev)	60.0±1.9	–	38.2
Fission neutrons (0.9 MeV)	72.8±2.4	–	46.4
Fission neutrons (0.7 MeV)	83.5±1.0	–	53.1

[a] The biological effect (E), or the number of dicentrics per cell, is given by $E=\alpha D+\beta D^2$, D being the absorbed dose. The maximum RBE at low doses, RBE$_M$, is then given by the ratio of the values of α.

Table 7. RBE values of fast neutrons, Co-60 gamma-rays and orthovoltage X-rays for the production of dicentric chromosomes in human lymphocytes. (After Fabry et al. 1985)

Radiation	RBE at a dose of 2.0 Gy gamma-rays	RBE at a dose of 0.5 Gy gamma-rays	RBE$_M$ α-test/α-ref
^{60}Co-gamma-rays	1	1	1
250-kV X-rays	1.2	1.3	1.5
Neutrons d(50)+Be	1.8	3.0	4.7
Neutrons d(33)+Be	2.3	3.8	6.2
Neutrons d(16)+Be	4.0	8.2	14.0

mean energy of 7.6 MeV. At Louvain-la-Neuve Fabry obtained RBE$_{max}$ values of 14, 6 and 5 (relative to gamma-rays) for neutrons produced by 16-, 33- and 50-MeV deuterons on a Be target (mean neutron energy: 6.5, 14 and 21 MeV, respectively).

Premature Chromosome Condensation

Premature chromosome condensation (PCC) assays have been used by Prasanna et al. (1997) to determine the RBE of fission neutrons (Y_D=65 keV/µm) compared to 250-kVp X-rays (Y_D=4 keV/µm). Human lymphocytes were irradiated in vitro, and a non-Poisson distribution was observed for prematurely condensed chromosomes and fragments, after repair at 37 °C for 24 h following exposure. From the dose-response relationships for the yield of excess PCC fragments, the RBE of fission neutrons relative to 250-kVp X-rays was estimated to be 2.4±0.39 (Fig. 16).

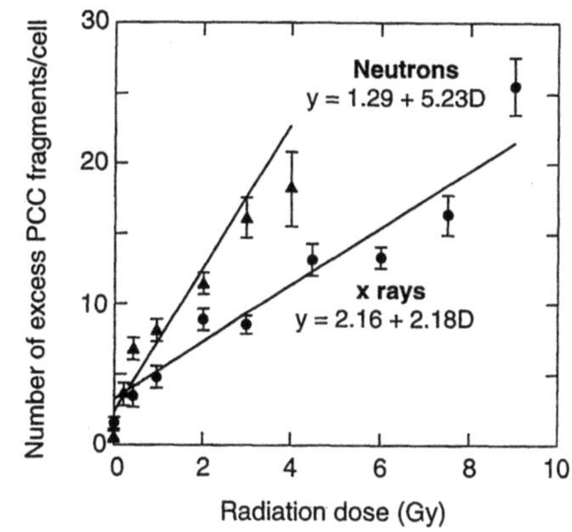

Fig. 16. Premature chromosome condensation (PCC) after irradiation with fission neutrons and 250 kVp X-rays. Dose-response relationships obtained for the yield of excess PCC fragments after 24 h repair at 37 °C. *Bars* represent SEs of the mean. (After Prasanna et al. 1997)

Neutron RBE for Genetic Effects

Neutron RBE for genetic effects has been studied in a variety of biological systems. The mechanisms of production of mutations are to a certain extent similar, allowing the results to be extrapolated cautiously to man (Hei et al. 1988).

In plant systems, induction of somatic mutations in stamen hair in *Tradescantia* has been studied by several authors (Sparrow et al. 1972; Underbrink et al. 1975, 1976; Dennis 1976; Pihet 1989). This biological system exhibits a high radiosensitivity and allows the study of the initial part of the dose-effect relationship.

Fission Neutrons

For fission neutrons, high RBE values have been observed. For example, for pink mutations in *Tradescantia*, Underbrink et al. (1976) observed an RBE of 100 for fission neutrons compared with Co-60 gamma-rays and 50 when compared with 200 kV X-rays.

Grahn et al. (1983) studied various effects in mice (dominant lethal mutations, chromosomal translocations and other abnormalities in spermatogonia). They found RBE_{max} values of 10–45 for fission neutrons compared with X-rays, depending on the biological criterion. Higher RBE values have been seen for chromosomal translocations in spermatogonia and for neutrons given once per week compared with gamma-rays at low dose rate.

Fast Neutrons

For fast neutrons, Pihet et al. (1985; Pihet 1989) found lower RBE values than those reported for fission neutrons. In comparison with gamma-rays, the RBE of neutrons produced by 65-MeV protons [p(65)+Be] was found to be

12±4 at a neutron dose of 4 mGy. Figure 17 a, b compares the results by Pihet (1989) for genetic effects with those of Underbrink and Dennis (1976), respectively.

Figure 18 summarizes the relations between RBE_{max} and neutron energy for induction of pink mutations in *Tradescantia*. For comparison, the data obtained for growth inhibition in *Vicia faba* (another criterion on a vegetal system) are shown on the same figure (Hall et al. 1973 and Van Dam 1984; both cited in Pihet 1989). The variations of the microdosimetric quantities yD and y^* are also shown for comparison.

As pointed out several times previously in this chapter, and at any neutron energy, the RBE for induction of mutations is significantly higher than the RBE for growth delay, which can be considered a consequence of a cell-killing effect.

Cataracts After Fission Neutron Irradiation

For production of cataracts, high RBE values have been obtained by Bateman et al. (1972), who compared fission neutrons (0.43 MeV) with 200-kV X-rays. RBE values greater than 100 were seen at 1 mGy neutrons and, in addition, the RBE was found to increase in proportion to $\sqrt{\frac{1}{D_n}}$ for doses at least as

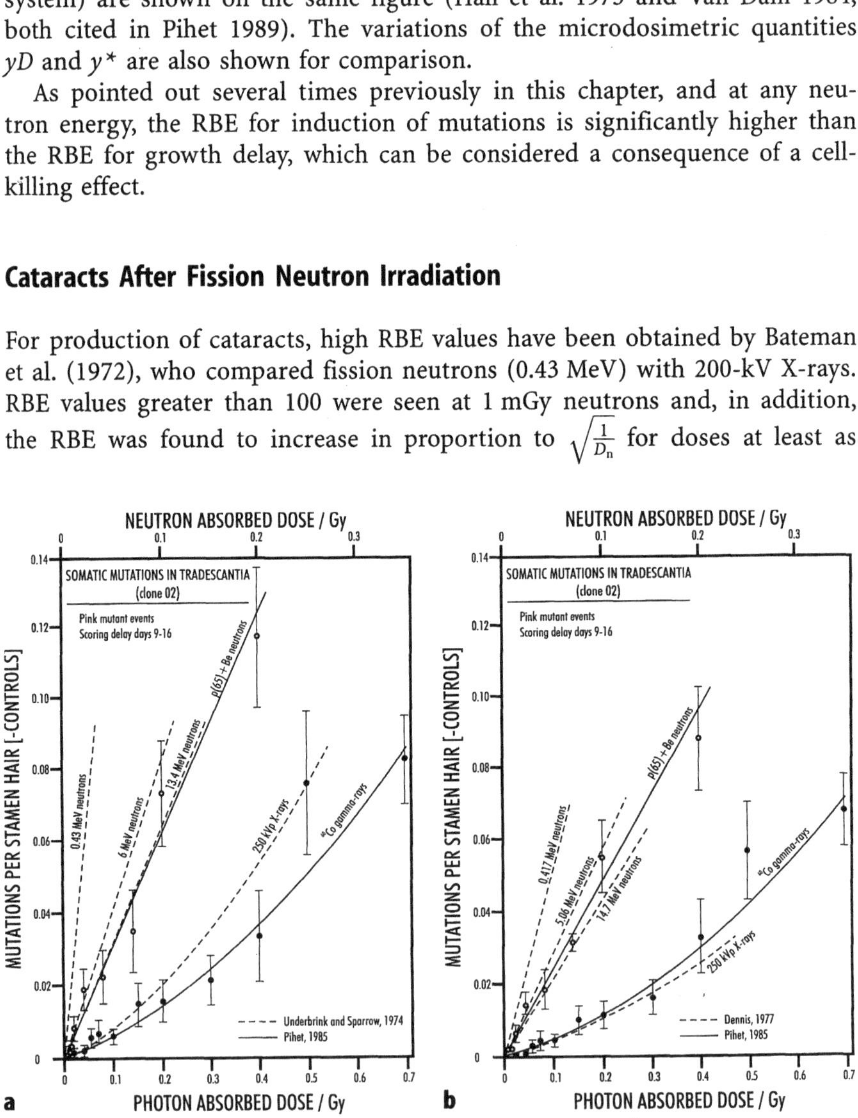

Fig. 17 a, b. Dose effect relations for induction of pink mutations in *Tradescantia* stamen hair. Two biological endpoints are selected: **a**, average number of mutations per stamen hair; **b** fraction of stamen hair with at least one mutation. The photon doses and the neutron doses are indicated *in the lower and upper abscissa*, respectively. (From Pihet 1989)

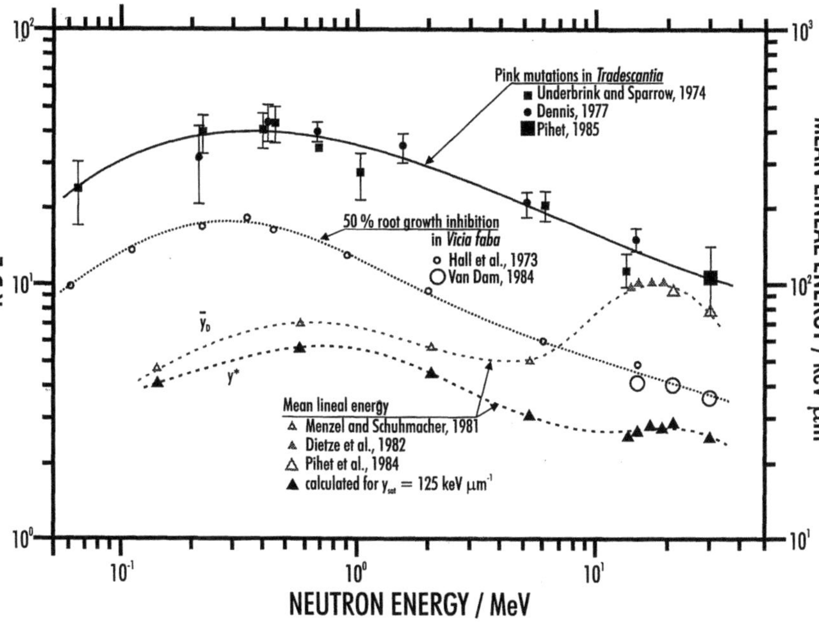

Fig. 18. Variation of RBE_{max} of neutrons as function of energy for different biological endpoints in vegetal systems: pink mutations in *Tradescantia* stamen hair and growth inhibition in *Vicia faba* roots. For comparison (*right ordinate*), the variation of the microdosimetric quantity mean linear energy: *yD* and *y**. (From Pihet 1989)

low as 1 mGy. These studies have been repeated by Di Paola et al. (1980), who found values of RBE from 25 to 40 at 10 mGy neutrons.

RBE of Alpha-Particles and Heavy Ions

Development of therapy with helion- and heavy-ion beams in several centres worldwide requires accurate information on RBE for the effects relevant to the therapy applications and related mainly to cell killing. On the other hand, for evaluation of the risks of the applications of these new beams, information is also needed on the RBE, at low doses, for cancer induction and effects resulting from genomic lesions.

Alpha-Particles

Limited data are available so far concerning the RBE of alpha-particles. They were reviewed by Sinclair (1985) and are presented in Table 8; most were obtained for biological criteria which are directly or indirectly relevant to the radiation protection situations. These limited data suggest that alpha-particles have RBEs which differ little from fission neutrons. The RBEs of both of

Table 8. RBE values for alpha-particles relative to fractionated gamma-rays or fission neutrons. (After Sinclair 1985)

	Gamma	n
Chromosome translocations in mouse spermatogonia, ^{239}Pu (Grahn et al. 1984)		1.0
Chromosome/chromatid fragments in mouse spermatogonia, ^{239}Pu (Grahn et al. 1984)		1.5
Mutations (*HGPRT* locus) in human cells, ^{238}Pu (Thacker et al. 1982) (10.3 for acute X-rays), estimate	20–30	
Chromosome aberrations		
Edwards et al. (1982)		~0.5
Vulpis (1973)		~1
Dufrain et al. (1979)		~1
Cancer in dogs, ^{238}Pu vs. ^{144}Ce (Hahn et al. 1984)	~25	
Bone cancer in dogs, ^{90}Sr vs. ^{226}Ra (Goldman et al. 1985)	25–50	

these beam qualities may thus well be 20–30 or higher relative to gamma-rays.

Induction of lung cancers after exposure to alpha-particles raises difficult practical problems and complex problems of interpretation. Therefore major discrepancies are often observed. Induction of pulmonary tumours after exposure to alpha-particles has been studied in mice by Lundgren et al. (1987). For a given total dose, a single inhalation of Pu-239 oxide was about four times more effective than repeated exposures. On the other hand, Little et al. (1985) performed intratracheal instillation of Po-210 in Syrian hamsters. Protraction of the alpha-irradiation over 120 days was slightly more carcinogenic at lower total lung doses but slightly less carcinogenic at higher doses, in comparison with an exposure limited to a 10-day period. However, the possible influence of other factors was illustrated by the fact that subsequent weekly instillation of saline alone markedly enhanced the carcinogenic effect of a single intratracheal instillation of Po-210.

Heavy Ions

Different heavy ions have been used in therapy (C, Ne) or for radiobiological experiments (C, Ne, ^{56}Fe, ^{40}Ar, etc.). Depending on their specific character and energy, they could have LET, or y, values close to the maximum for neutrons, or beyond the peak of the RBE vs. y curve, especially for heavier and more energetic ions. Figure 19 compared the microdosimetric distributions measured for ^{60}Co, protons, nitrogen, neon, argon and iron ions (Dicello et al. 1991). At equal doses, there are fewer events with larger energy depositions per event when irradiating with heavy ions. A large number of radiobiological studies were performed with heavy ions, but mainly using non-stochastic endpoints, for therapy applications. The results of some of the studies performed, at low doses, for endpoints relevant in radiation protection are summarized in Table 9.

Inactivation and mutation to thioguanine resistance were studied in cultured V79 Chinese hamster cells and in human diploid fibroblasts after irradiation with a variety of accelerated helium, boron and nitrogen ions over a range of LETs from 20 keV/μm to 470 keV/μm (Cox et al. 1977). The RBEs for mutation peaked in the range 90–200 keV/μm and were about twice as large as the RBEs for inactivation. The RBEs were about 8–12 compared to acute 250-kVp X-rays, and therefore presumably would be ~30 compared to protracted gamma-radiation (Sinclair 1985).

For the induction of tumours in the Harderian gland of the mouse, the same RBE was observed, i.e. RBE ~25 at 0.1 Gy (20% tumour prevalence), for both ^{56}Fe ions (180 keV/μm) and ^{40}Ar ions (650 keV/μm) as for fission neutrons.

For cataract induction by ^{40}Ar ions, relative to X-rays over a wide range of doses, at the lowest significant dose of 0.05 Gy of ^{40}Ar ions, the RBE was 40. The RBE vs. dose relationship followed the equation RBE $= C\sqrt{D}$, with $C=8$ Gy$^{1/2}$ for the ^{40}Ar ions compared with 5 Gy$^{1/2}$ for fission neutrons. ^{40}Ar ions may thus have a higher RBE than fission neutrons for cataract induction.

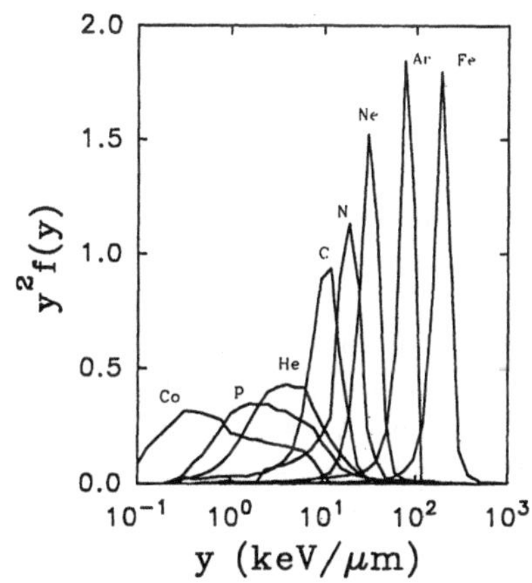

Fig. 19. Comparison of microdosimetric distributions measured for ^{60}Co, protons and energetic heavy ion beams of helium, carbon, nitrogen, neon, argon and iron. In this representation, the total dose is proportional to the area under each curve and is the same for each type of particle. (After Dicello et al. 1991)

Table 9. RBEs of heavy ions relative to fractionated gamma-rays. (After Sinclair 1985)

Mutations (*HGPRT* locus) in humans cells, 90–200 keV/μm (Cox et al. 1977) (8–12 for acute X-rays), estimate	20–30
Tumours in Harderian gland (mouse) (Fry et al. 1983), ^{56}Fe and ^{40}Ar	~25
Cataract, ^{40}Ar ions (Merriam et al. 1984)	40

It can be concluded that the maximum value of the RBE for heavy ions for carcinogenesis and mutagenesis is probably not substantially different from that of neutrons with optimum energy or alpha-particles. At lower and perhaps at higher LET, the RBE of heavy ions could be expected to be less.

Conclusions

In order to evaluate the risk of exposure to neutron irradiation, besides the dosimetric parameters, information is needed about the RBE. Since at present no human data are available concerning the RBE of neutrons for cancer induction, animal data have to be relied upon; of course, extrapolation of animal data to the human situation should be done with the greatest caution.

The RBEs of neutrons, relative to X- or gamma-rays, depend on several factors, first of all on the biological system and endpoint, and the radiation dose level. Therefore, the RBEs are in general higher for effects relevant in radiation protection (such as cancer induction, chromosome aberrations, genetic effects) than for effects relevant in radiation therapy (and related mainly to cell killing).

The RBE of a neutron beam also depends on the neutron energy spectrum (which is related to LET and microdosimetric spectrum (ICRU 1977, 1980, 1993b). This explains the importance and the potential role of microdosimetry, from a practical and/or operational point of view, in both the fields of radiation therapy and protection (ICRU 1983).

Lastly, when reporting a neutron RBE value, allowance must be made for the reference radiation quality since there is evidence that the RBE of 200-kV X-rays could be as high as 2 compared with cobalt-60 gamma-rays at low doses and low dose rates.

For neutrons with energies of 0.43–1 MeV, such as those produced by nuclear fission, a large range of RBE values has been reported (e.g. from 3 to 200 for tumour induction), depending on the biological criterion and, for a given criterion, from one centre to another, depending on the experimental conditions. A careful and comprehensive review was done by Sinclair in 1985, and few new data have been published since. As can be seen in Table 10, a large proportion of the RBE values are in the range of 15–70.

The neutron RBE decreases at higher energies, and thus the RBEs of fast neutrons used in radiation therapy are lower than the RBEs of fission neutrons. The RBEs of fast neutrons can be estimated to be three to four times lower than for fission neutrons (Edwards 1997). This can be explained from the shape of the y spectra (or LET spectra). As illustrated in Fig. 5, the y spectra for the neutron beams used in therapy extend over more than four orders of magnitude. Part of the energy is delivered at lower y (or LET) than the fission neutrons, thus with a lower RBE. In contrast, the other part of the energy is delivered at high y (or LET), in a y (or LET) range where the RBE decreases due to the "overkill effect".

Table 10. RBEmax value of fission neutrons, relative to gamma-rays after fractionated exposure (summary values). (After Sinclair 1985)

Effect	RBE_M
Tumor induction	≈3 to ≈200
Life shortening	15–45
Cell transformation	35–70
Chromosome aberrations	40–50
Genetic effects in mammals	10–45
Cataracts	25–200
Micronuclei	6–60
Loss of weight of the testis	5–20

In addition to the neutron RBE values obtained for tumour induction, the available RBE values for other effects, which could be to some extent related to tumour induction, have been reviewed in this chapter. For example, it is recognized that, at low dose, life shortening is due mainly to cancer induction, and thus an evaluation of life shortening can be an indirect measure of cancer induction. Evaluation of other effects such as malignant cell transformation, chromosome aberrations, and genetic effects, which express to some extent "genomic damage", can also contribute to support and validate the conclusions derived directly from observation of cancer induction in animals.

In radiation protection, in order to take into account the fact that, at equal dose, the probability of inducing a stochastic effect (e.g. cancer induction or hereditary effect) depends on the radiation quality, the ICRP has introduced the quantity of *radiation weighting factor*, W_R (ICRP 1991, 1996).

Following the ICRP recommendations, the probability of inducing a stochastic effect in an organ depends on the product of the absorbed dose averaged over that organ, and a weighting factor for the radiation quality (W_R) incident of the body. This weighted absorbed dose is the *equivalent dose* H_T for that organ (or tissue). The equivalent dose in organ/tissue T is thus given by the expression:

$$H_T = W_R \cdot D_R$$

If different radiation qualities (with different W_R) are involved, the corresponding H_T values must be added.

The values of the radiation weighting factors for the different radiation qualities (type and energy) have been selected by the ICRP to be representative of the RBE values of these radiations in inducing stochastic effects at low doses. The radiation weighting factors W_R recommended by the ICRP are given in Table 11 and in Fig. 20. It can be concluded from the present review of the available radiobiological data that these W_R values for neutrons are a reasonable compromise for the purpose of risk assessment and limitation in occupational and public exposures.

Table 11. Radiation weighting factors W_R. (ICRP Publication 1991)

Type and energy range[a]	Radiation weighting factor, W_R
Phtoons, all energies	1
Electrons and muons, all energies[b]	1
Neutrons, energy <10 keV	5
10 keV to 100 keV	10
>100 keV to 2 MeV	20
>2 MeV to 20 MeV	10
>20 MeV	5
Protons, other than recoil protons, energy >2 MeV	5
Alpha-particles, fission fragments, heavy nuclei	20

[a] All values relate to the radiation incident on the body or, for internal sources, emitted from the source.

[b] Excluding Auger electrons emitted from nuclei bound to DNA.

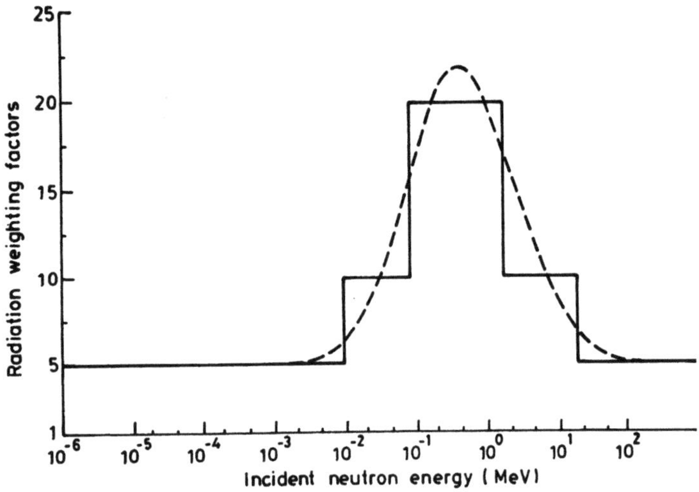

Fig. 20. Radiation weighting factors W_R for neutrons. The smooth curve is to be treated as an approximation. (From ICRP 1991)

In radiation therapy, the risk of inducing a cancer by the irradiation can of course not be ignored, although it is obvious that is is in general small compared to the risk of dying from the first cancer (or lesion) for which the treatment was initially prescribed. A similar problem is raised by the combination of radiations with drugs, some of these drugs being particularly harmful as fas as secondary cancer induction is concerned.

As far as introduction of high-LET radiations is concerned, it is true that the RBE for tumour induction is significanlty higher than for low-LET radiations. Here again the risk of tumour induction has to be balanced against the expected improvement in cure rate when the indication for high-LET radiation therapy has been correctly evaluated in well-selected patient groups (Wambersie 1992).

As far as carbon ions as used in radiation therapy are concerned, the available data indicate that their average RBE is comparable to the RBE of fission neutrons. However, for carbon ion beams, the high physical selectivity of the irradiation significantly reduces the volume of the irradiated normal tissues and thus the risk of cancer induction. The integral doses are indeed about 30% of that obtained with fast neutron and photon beams, depending of course on the tumour site. The same arguments are valid for the other heavy ion beams used in therapy.

In conventional photon beam therapy, the small proportion of neutrons which are indeed present in the photon beams has been a source of concern (when their energy exceeds about 10 MV) because of the increased risk of cancer induction by high-LET radiations. Here again this type of risk has to be balanced against the benefit of an improved dose distribution when increasing photon beam energy. The dose distribution is generally considered as the main factor for selecting the photon energy, but this may be (and has been) questioned.

Lastly, for safety, high-LET radiations should in general not be used for treatment of children, at least when a high probability of cure is expected, because of increased risk of secondary cancer induction.

References

Barendsen GW (1968) Responses of cultured cells, tumours and normal tissues to radiations of different linear energy transfer. Curr Top Radiat Res Q 4:293–356

Barendsen GW (1979) Influence of radiation quality on the effectiveness of small doses for induction of reproductive death and chromosome aberrations in mammalian cells. Int J Radiat Biol 36:49–63

Barendsen GW (1996) RBE-LET relationships for lethal, potentially lethal and sublethal damage in mammalian cells: implications for fast neutron radiotherapy. Bull Cancer (Paris) 83 Suppl 1:15 s–18 s

Bateman JL, Rossi HH, Kellerer AM, Robinson CV, Bond VP (1972) Dose-dependence of fast neutron RBE for lens opacification in mice. Radiat Res 51:381–390

Bauchinger M, Koester L, Schmid E, Dresp J, Streng S (1984) Chromosome aberrations in human lymphocytes induced by fission neutrons. Int J Radiat Biol 45:449–457

Boice JD Jr, Engholm G, Kleinerman RA et al (1988) Radiation dose and second cancer risk in patients treated for cancer of the cervix. Radiat Res 116:3–55

Breteau N, Le Bourgeois J-P, Barendsen GW, Stannard CE, Rosenwald J-C, Wambersie A (1996) Hadrons in radiation therapy. Bull Cancer (Paris) 83 Suppl

Broerse JJ, Hennen LA, van Zwieten MJ (1985) Radiation carcinogenesis in experimental animals and its implications for radiation protection. Int J Radiat Biol 48:167–187

Broerse JJ, Van Bekkum DW, Zurcher C (1989) Radiation carcinogenesis in experimental animals. Experientia 45:60–69

Cox R, Thacker J, Goodhead DT (1977) Inactivation and mutation of cultured mammalian cells by aluminium characteristic ultrasoft X-rays. II. Dose-responses of Chinese hamster and human diploid cells to aluminium X-rays and radiations of different LET. Int J Radiat Biol 31:561–576

Curtis RE, Boice JD Jr, Stovall M et al (1992) Risk of leukemia after chemotherapy and radiation treatment for breast cancer. N Engl J Med 326:1745–1751
Day NE, Boice JD Jr (eds) Second cancers in relation to radiation treatment for cervical cancer. IARC Sci Publ 52
De Vathaire F, Francois P, Hill C et al (1989) Role of radiotherapy and chemotherapy in the risk of second malignant neoplasms after cancer in childhood. Br J Cancer 59:792–796
Dennis JA (1976) Somatic aberrations induction in Tradescantia occidentalis by neutrons, X and gamma radiations. II. Biological results, RBE and OER. Int J Radiat Biol 29:323–342
Di Paola M, Coppola M, Baarli J, Bianchi M, Sullivan AH (1980) Biological responses to various neutron energies from 1 to 600 MeV. II. Lens opacification in mice. Radiat Res 84:453–461
Dicello Jf, Wasiolek M, Zaider M (1991) Measured microdosimetric spectra of energetic ion beams of Fe, Ar, Ne and C: limitations of LET distributions and quality factor in radiation effects and space research. IEEE Trans Nucl Sci 38:1203–1209
Dobson RL, Straume T (1982) Cancer risk and neutron RBE's from Hiroshima and Nagasaki. In: Broerse JJ, Gerber GB (eds) Neutron carcinogenesis. Commission of the European Communities, Luxembourg, pp 279–299. Radiation protection report EUR 8084 EN
Dufrain RJ, Littlefield LG, Joiner EE, Frome EL (1979) Human cytogenetic dosimetry: a dose response relationship for alpha radiation from 241 Am. Health Phys 37:279–290
Edwards AA (1997) RBE values for neutron radiations. National Radiological Protection Board, Chilton, pp 7–8, Bulletin 193
Edwards AA, Lloyd DC, Purrott RJ, Prosser JS (1982) The dependence of chromosome aberration yields on dose rate and radiation quality. In: Research and development report, 1979–1981. National Radiological Protection Board, Chilton, pp 83–85
Fabry L, Leonard A, Wambersie A (1985) Induction of chromosome aberrations in Go human lymphocytes by low doses of ionizing radiations of different quality. Radiat Res 103:122–134
Fry RJM, Powers-Risius P, Alpen EL, Ainsworth EJ, Ullrich RL (1983) High LET radiation carcinogenesis. Adv Space Res 3(8):241–243
Furuse T, Noda Y, Otsu H, Ohara H (1997) High-LET induced tumours in mice: tumour spectrum and RBEs. In: National Institute of Radiological Sciences (ed) Annual report 1996. NIRS, Chiba, pp 48–49
Geard CR (1996) Neutron induced recoil protons of restricted energy and range and biological effectiveness. Health Phys 70(6):804–811
Geard CR, Brenner DJ (1997) Chromosome aberrations induced by neutrons: implications for radiation protection. In: EDF (ed) Effects biologiques des neutrons: conséquences en radioprotection. Electricité de France, Paris, pp 31–43 (Report no 10)
Grahn D, Fritz T (1984) Studies on chronic radiation injury with mice and dogs exposed to external whole body irradiation at ANL. In: Proceedings of the 22nd Symposium on Life Sciences, Batelle Pacific Northwest Laboratory, Hanford, 27–30 Sept 1983: US DOE Conf-830951
Grahn D, Thomson JF, Williamson FS, Lombard LS (1983) Somatic and genetic effects of low doses of fission neutrons and Co-60 gamma rays. In: Proceedings of the 7th International Congress of Radiation Research, Amsterdam 3–10 July 1983. Abstract Book C. Nijhoff, Dordrecht, pp C2–05
Gueulette J, Beauduin M, Grégoire V, Vynckier S, De Coster BM, Octave-Prignot M, Wambersie A, Strijkmans K, De Schrijver A, El-Akkad S, Böhm L, Slabbert JP, Jones DTL, Maughan R, Onoda J, Yudelev M, Porter AT, Powers WE, Sabattier R, Breteau N, Courdi A, Brassart N, Chauvel P (1996) RBE variation between fast neutron beams as a function of energy. Intercomparison involving 7 neutrontherapy facilities. Bull Cancer (Paris) 83 Suppl 1:55 s–63 s
Hahn A, Hill CK, Elkind MM (1984) Repair processes and radiation quality in neoplastic transformation of mammalian cells. Radiat Res 199:249–261
Hall EJ (1996) Neutrons and carcinogenesis: a cautionary tale. Bull Cancer (Paris) 83 Suppl 1:43 s–46 s

Hall EJ, Novak JK, Kellerer AM, Rossi HH, Marino S, Goodman LJ (1975) RBE as function of neutron energy. I. Experimental observations. Radiat Res 64:245–255
Hei TK, Hall EJ, Waldren CA (1988) Mutation induction and relative biological effectiveness of neutrons in mammalian cells. Radiat Res 115:281–291
Hempelmann L, Lushbaugh C, Voelz G (1980) What happened to the survivors of the early Los Alamos nuclear accidents? In: Hubner K, Fry S (eds) The medical basis for radiation accident preparedness. Elsevier/North-Holland, New York, pp 17–32
Hill CK, Carnes BA, Han A, Elkind MM (1985) Neoplastic transformation is enhanced by multiple low doses of fission-spectrum neutrons. Radiat Res 102:404–410
Hollander CF, van Zwieten MJ, Broerse JJ (1982) Carcinogenesis studies in rhesus monkeys after fission neutron and X-irradiation. In: Broerse JJ, Gerber GB (eds) Neutron carcinogenesis. Commission of the European Communities, Luxembourg, pp 183–185 (Radiation protection report EUR 8084 EN)
ICRP (1963) Report of the RBE Subcommittee to the International Commission of Radiological Protection and the International Commission on Radiation Units and Measurements. Health Phys 9:357–386
ICRP (1991) 1990 recommendations of the International Commission on Radiological Protection. Pergamon, Oxford (ICRP publ 60)
ICRP (1996) Conversion coefficients for use in radiological protection against external radiation. Report of a Joint Task Group of the International Commission on Radiological Protection (ICRP) and of the International Commission on Radiation Units and Measurements (ICRU). Pergamon, Oxford (ICRP publ 74, vol 26)
ICRU (1977) Neutron dosimetry for biology and medicine. International Commission on Radiation Units and Measurements, Bethesda (ICRU report 26)
ICRU (1978) Dose specification for reporting external beam therapy with photons and electrons. International Commission on Radiation Units and Measurements, (ICRU report 29)
ICRU (1980) Radiation quantities and units. International Commission on Radiation Units and Measurements, Bethesda (ICRU report 33)
ICRU (1983) Microdosimetry. International Commission on Radiation Units and Measurements, Bethesda (ICRU report 36)
ICRU (1992) Photon, electron, proton and neutron interaction data for body tissues. International Commission on Radiation Units and Measurements, Bethesda (ICRU report 46)
ICRU (1993a) Stopping powers and ranges for proton and alpha particles. International Commission on Radiation Units and Measurements, Bethesda (ICRU report 49)
ICRU (1993b) Quantities and units in radiation protection dosimetry. International Commission on Radiation Units and Measurements, Bethesda (ICRU report 51)
Karzmark CJ (ed) (1987) Total skin electron therapy: technique and dosimetry. American Institute of Physics, New York (American Association of Physicists in Medicine, report 23)
Karzmark CJ, Nunan S, Tanabe E (1993) Medical electron accelerators. McGraw-Hill, New York
Kellerer AM, Rossi HH (1972) The theory of dual radiation action. Curr Top Radiat Res Q 8:85
Kliauga PJ, Dvorak R (1978) Microdosimetric measurements of ionization by monoenergetic photons. Radiat Res 73:1
Léonard A, Gerber GB (1982) RBE and dose-effect relationships in mammalian somatic and germ cells. In: Broerse JJ, Gerber GB (eds) Neutron carcinogenesis. Commission of the European Communities, Luxembourg, pp 361–364 (Radiation protection, report EUR 8084 EN)
Little JB, Kennedy AR, McGandy RB (1985) Effect of dose rate on the induction of experimental lung cancer in hamsters by alpha radiation. Radiat, Res 103:293–299
Lundgren DL, Gillet NA, Hahn FF, Griffith WC, McClellan RO (1987) Effects of protraction of the alpha dose to the lungs of mice by repeated inhalation exposure to aerosols of Pu-239-oxide. Radiat Res 111:201–224
Maisin JR, Wambersie A, Gerber GB, Gueulette J, Mattelin G, Lambiet-Collier M (1983) Life shortening and disease incidence in BALB/c mice following a single d(50)+Be neutron or gamma exposure. Radiat Res 94:374–389

Maisin JR, Wambersie A, Gerber GB, Mattelin G, Lambiet-Collier M, De Coster B, Gueulette J (1988) Life shortening and disease incidence in C57Bl mice after single and fractionated γ and high-energy neutron exposure. Radiat Res 113:300–317

Maisin JR, Vankerkom J, De Saint-Georges L, Janowski M, Lambiet-Collier M, Mattelin G, Wambersie A (1995) Effects of neutrons alone or combined with diethylnitrosamine on tumour induction in the livers of infant C57BL mice. Radiat Res 142:78–84

Maisin JR, Gerber GB, Vankerkom J, Wambersie A (1996) Survival and diseases in C57BL mice exposed to X rays or 3.1 MeV neutrons at an age of 7 or 21 days. Radiat Res 146:453–460

Merriam GR, Worgul BV, Medvedovsky C, Zaider M, Rossi HH (1984) Accelerated heavy particles and the lens. I. Cataractogenic potential. Radiat Res 98:129–140

Nath R, Epp ER, Laughlin JS, Swanson WP, Bond VP (1984) Neutrons from high-energy x-ray medical accelerators: an estimate of risk to the radiotherapy patient. Med Phys 11:231–241

NCRP (1990) The relative biological effectiveness of radiations of different quality. National Council on Radiation Protection and Measurements. Bethesda (Report no 104)

Pihet P (1989) Etude microdosimétrique des faisceaux de neutrons de haute énergie. Applications dosimétriques et radiobiologiques. PhD thesis, Catholic University of Louvain

Pihet P, Menzel HG, Haut H, Garot PJ, Wambersie A (1985) Shape of the RBE dose relationship for fast neutrons at low doses for somatic mutations in Tradescantia. In: Schraube H, Burger G (eds) Proceedings of the 5fth Symposium on Neutron Dosimetry, Munich 1984, vol 1. Commission of the European Communities, Luxembourg, pp 45–55 (Report no EUR 9762)

Pohl-Rüling J, Fischer P, Lloyd DC, Edwards AA, Natarajan At et al (1986) Chromosomal damage in human lymphocytes by low doses of D-T neutrons. Mutat Res 173:267–272

Prasanna PGS, Kolanko CJ, Gerstenberg HM, Blakely WF (1997) Premature chromosome condensation assay for biodosimetry: studies with fission neutrons. Health Phys 72:594–600

Rodgers RC, Gross W (1974) Microdosimetry of monoenergetic neutrons. In: Proceedings of the 4th Symposium on Microdosimetry. Booz J, Ebert H, Eickel R, Waker A (eds) Commission of the European Communities, Luxembourg, p 1027 (Report no EUR 5122)

Rossi HH (1996) Radiation physics and radiobiology. Health Phys 70(6):828–831

Sabatier L, Al Achkar W, Hoffschir F, Luccioni C, Dutrillux B (1987) Qualitative study of chromosomal lesions induced by neutrons and neon ions in human lymphocytes at G_O phase. Mutat Res 178:91–97

Savage JRK (1982) RBE of neutrons for genetic effects. In: Broerse JJ, Gerber GB (eds) Neutron carcinogenesis. Commission of the European Communities, Luxembourg, pp 307–332 (Radiation protection, report EUR 8084 EN)

Savage JRK, Holloway M (1988) Induction of sister-chromatid exchanges by d(42 MeV)-Be neutrons in unstimulated human-blood lymphocytes. Br J Radiol 61:231–234

Shellabarger CJ, Chmelevsky D, Kellerer AM (1980) Induction of mammary neoplasms in the Sprague-Dawley rat by 430 keV neutrons and X-rays. J Natl Cancer Inst 64:821–833

Shellabarger CJ, Chmelevsky D, Kellerer AM, Stone JP, Holtzman S (1982) Induction of mammary neoplasms in the ACI rat by 430 keV neutrons, X-rays and diethylstilboestrol. J Natl Cancer Inst 69:1135–1146

Sinclair WK (1985) Experimental RBE values of high LET radiations at low doses and the implications for quality factor assigment. Radiat Protect Dosimetry 13 (1–4):319–326

Sinclair WK (1996) The present system of quantities and units for radiation protection. Health Phys 70(6):781–786

Sparrow AH, Underbrink AG, Rossi HH (1972) Mutations induced in Tradescantia by small doses of X-rays and neutrons: analysis of dose-response curve. Science 76:916–918

Storer JB, Mitchell TJ (1984) Limiting values for the RBE of fission neutrons at low doses for life shortening in mice. Radiat Res 97:396–406

Straume TS, Egbert SD, Woolson WA, Finckel RC, Kubik PW, Gove HE, Sharma P, Hoshi M (1992) Neutron discrepancies in the DS86 Hiroshima dosimetry system. Health Phys 63(4):421–426

Thacker J, Stretch A, Goodhead DT (1982) The mutagenecity of alpha particles from 239 Plutonium. Radiat Res 92:343–352

Thomson JF, Williamson FS, Grahn D (1985) Life shortening in mice exposed to fission neutrons and gamma-rays. Radiat Res 104:420–428

Tucker MA, Coleman CN, Cox RS et al (1988) Risk of second cancers after treatment for Hodgkin's disease. N Engl J Med 318(2):76–81

Tubiana M, Dutreix J, Wambersie A (1990) Neutrons and other heavy particles. In: Introduction to radiobiology. Taylor and Francis, London, pp 273–312

Ullrich RL (1982) Lung tumor induction in mice: neutron RBE at low doses. In: Broerse JJ, Gerber GB (eds) Neutron carcinogenesis. Commission of the European Communities, Luxembourg, pp 43–55 (Radiation protection, report EUR 8084 EN)

Ullrich RL (1983) Tumor induction in BALB/c female mice after fission neutron or gamma-irradiation. Radiat Res 93:506–515

Ullrich RL (1984) Tumor induction in BALB/c mice after fractionated and protracted exposures to fission-spectrum neutrons. Radiat Res 97:587–597

Underbrink AG, Sparrow AH, Sautkulis D, Mills RE (1975) An elusive factor affecting mutation frequency in Tradescantia stamen hairs: its influence on RBE. Int J Radial Biol 28:527–538

Underbrink AG, Kellerer AM, Mills RE, Sparrow AH (1976) Comparison of x-ray and gamma-ray dose responses curves for pink somatic mutations in Tradescantia clone 02. Radiat Environ Biophys 13:295–303

UNSCEAR (1982) Ionizing radiation: sources and biological effects. United Nations Scientific Committee on the Effects of Atomic Radiations, New York (1982 report to the General Assembly, with annexes)

UNSCEAR (1993) Report to the General Assembly, with scientific annexes. United Nations Scientific Committee on the Effects of Atomic Radiations, New York (United Nations publication sales no E.94.IX.2)

Van Dam J, Bonte J, Dutreix A, Bouhnik H, Wambersie A (1982) RBE of californium-252 at low dose rate for growth inhibition in Vivia faba. J Eur Radiother 3:109–120

Van der Kogel AJ (1979) Late effects of radiation on the spinal cord: dose-effect relationships and pathogenesis. Thesis, University of Amsterdam

Vogel HH, Dickson HW (1982) Mammary neoplasia in Sprague-Dawley rats following acute and protracted irradiation. In: Broerse JJ, Gerber GB (eds) Neutron carcinogenesis. Commission of the European Communities, Luxembourg, pp 135–168 (Radiation protection, report EUR 8084 EN)

Vulpis N (1973) Chromosome aberrations induced in human peripheral blood lymphocytes using heavy particles from the 10B(n,alpha)Li reaction. Mutat Res 18:103–111

Wambersie A (1992) Neutron therapy: from radiobiological expectation to clinical reality. Radiat Protect Dosimetry 44:379–395

Wambersie A, Gahbauer RA (1996) Medical applications of electron linear accelerators. In: Turner S (ed) CERN Accelerator School (CAS): Cyclotrons, linacs and their applications. CERN 96-02, 4 March 1996. CERN, Geneva, pp 229–248

Wambersie A, Menzel HG (1997) Specification of absorbed dose and radiation quality in heavy particle therapy (a review). Radiat Protect Dosimetry 70:517–527

Wambersie A, Grégoire V, Brucher J-M (1992) Potential clinical gain of proton (and heavy ion) beams for brain tumors in children. Int J Radiat Oncol Biol Phys 22:275–286

Wolber G, Guibbaud Y, Dollo R (1996) Neutron dosimetry in French nuclear power plants. Problems and their solutions in 1995. Bull Cancer (Paris) 83 Suppl 1:19 s–26 s

Neutron Therapy in Malignant Salivary Gland Tumors: Results at European Centers

A. Krüll[1], R. Schwarz[1], S. Brackrock[1], R. Engenhart-Cabillic[2], P. Huber[2], F. J. Prott[3], N. Breteau[4], A. Favre[4], A. Lessel[5], H. Koppe[5], and T. Auberger[6]

[1] Abteilung für Strahlentherapie, Universitätskrankenhaus Hamburg-Eppendorf, Martinistrasse 52, D-20246 Hamburg, Germany
[2] Abteilung für Strahlentherapie, Medizinisches Zentrum für Radiologie, Baldingerstrasse, D-35033 Marbug, Germany
[3] Klinik und Poliklinik für Strahlentherapie und Radioonkologie, Westfälische Wilhelms-Universität, Albert-Schweitzer-Strasse 33, D-48129 Münster, Germany
[4] Service d'Oncologie et de Radiothérapie, Centre Hospitalier Régional d'Oreléans, Hôpital de la Source, BP 67009, F-45067 Orléans cedex 2, France
[5] Virchow-Klinikum, Robert-Rössle-Klinik, Funktionsbereich Strahlentherapie, Lindenberger Weg 80, D-13125 Berlin-Buch, Germany
[6] Universitätsklinik für Strahlentherapie, Leopold-Franzens-Universität, Anichstrasse 35, A-6020 Innsbruck, Austria

Abstract

Malignant salivary gland tumors are relatively rare and account for only 3%–4% of all head and neck cancers. Especially in adenoid cystic carcinoma, the incidence of local recurrence and distant metastases is influenced by the perineural spread of tumor. In advanced salivary gland tumors, surgery alone has a high incidence of local failure. The results of conventional radiotherapy are suboptimal in inoperable or not completely resected salivary gland tumors and in recurrent disease. The pooled data of some international series for low linear energy transfer radiation show a local control of only 28%. Especially in advanced salivary gland tumors, neutron therapy can improve local control. In Europe at least 570 patients with salivary gland tumors have been treated with neutrons alone or with combined modalities. The clinical data of different centers in Europe and the United States result in a similar local control rate of 67% in gross disease. An analysis of different European series shows on average a complication rate of 10.6% for severe radiation-related morbidity. Modern neutron machines and the use of three-dimensional treatment planning systems are now available in a few institutions and may further reduce side effects.

Introduction

Malignant salivary gland tumors are relatively rare and account for only 3%–4% of all malignancies of head and neck (Perez and Brady 1992). These tumors arise in major and minor salivary glands or in aberrant tissue (Hun-

Recent Results in Cancer Research, Vol. 150

ter et al. 1983). Most of the neoplasms occur in the parotid gland. Salivary gland tumors consist of different histologic types and subtypes which have varying potential for aggressive local behavior and metastasis (Laramore 1987; Seifert et al. 1986). Especially in adenoid cystic carcinoma, the incidence of local recurrence and distant metastasis is influenced by the perineural spread of tumor (Vrielinck et al. 1988). The mean age of patients with a malignant salivary gland tumor in 55 years (Million and Cassisi 1994). The etiologic factors are still under discussion. In Hiroshima an increased rate of malignancies was observed among the survivors of the atomic bomb (Takeichi et al. 1976). Nutrition has also been discussed as a factor (Perez et al. 1992).

In advanced salivary gland tumors surgery alone has a high incidence of local failure (Spiro 1986). Postoperative radiotherapy should be carried out after the indications have been carefully considered (Armstrong et al. 1990; Glanzmann et al. 1990; North et al. 1989; Rink and Becker 1990; Sadeghi et al. 1993; Shingaki et al. 1980; Wendt et al. 1992). Most patients with malignant salivary gland tumors in the world are irradiated with photons or electrons, but conventional radiotherapy has met with limited success in the setting of gross disease. For this reason salivary gland tumors have been a natural tumor system for neutron therapy studies and nowadays many authors prefer neutron therapy after an incomplete resection, in inoperable tumors or in recurrent disease (Buchholz et al. 1992; Catterall 1987; Duncan et al. 1987; Koh et al. 1989; Laramore et al. 1993; Pötter et al. 1990).

Materials and Methods

One of the first historical series in Europe was published by Eichhorn in 1981, who described the outcome of five patients with malignant salivary gland tumors irradiated at the neutron facility of Dresden-Rossendorf. Local control was achieved in three of these patients by neutron therapy.

Skolyszewski et al. (1982) published the data from three patients irradiated with neutrons at the Institute of Nuclear Physics in Krákow. In two control was achieved by high linear energy transfer (LET) radiation. One patient died during therapy with signs of general sepsis and cardiac failure.

In 1984 Catterall et al. (1984) reported data from 31 patients with advanced malignant tumors of the maxillary sinus, treated with fast neutrons at the Hammersmith Hospital in London. Treatment was given to 7 patients with adenocarcinomas, 6 patients with adenoid cystic carcinomas, 1 patient with a malignant melanoma, 12 patients with squamous cell carcinomas and 5 patients with transitional cell tumors. The target dose was 15.6 Gy applied in 12 fractions.

After neutron therapy, complete tumor remission was diagnosed in 29 patients, 4 of whom subsequently relapsed. The complete regression rate was 94% and the recurrence rate was 14%. The median survival time was 36 months. Complications were seen in ten patients, including severe side effects in most of them. Three of the ten patients had been irradiated previously. Catterall et al. (1984) explained the high incidence of complications as being due in part to the extent of tumor with destruction of normal tissue and also to the inadequacy of the used neutron beam.

Battermann and Mijnheer (1986) described 32 patients with inoperable salivary gland tumors who were treated with neutrons at the Netherlands Cancer Institute in Amsterdam. The majority of these patients suffered from a parotid gland tumor. The applied total dose varied between 15 and 22 Gy. Included were four patients with recurrent disease who had been previously irradiated with photons. The median survival time was 22 months. A complete tumor regression at 2 months was noticed in 25 patients and lasting local control could be observed in 21 cases. Serious complications developed in four patients, but all had previously been irradiated with photons.

Catterall published in 1987 data from 65 patients with locally advanced or recurrent malignant salivary gland tumors irradiated with neutrons between December 1969 and October 1984. Forty-four patients had tumors of the parotid gland, 9 had tumors of the submandibular gland, and other locations were involved in the remaining 12 patients. Fourteen of the 64 patients had previously been treated with photons. Fifty-eight tumors were classified as stage T_{4a} or T_{4b} and seven tumors were classified as stage T_2. The mean follow-up was 43 months. The local control of all patients with T_4 tumors was 76% and 86% in patients who had not previously been treated. Complete tumor regression occurred in all patients with a T_2 tumor, but one patient subsequently relapsed. For recurrent tumors local control was only 35% at 5 years. Radiation-induced complications were diagnosed in 16% of all previously untreated patients compared to 21% of patients who had previously undergone surgery or X-ray therapy. Of patients with parotid gland tumor, treated with neutrons after biopsy, only 77% maintained or regained facial nerve funtion.

Duncan et al. (1987) reported results from 28 patients with malignant salivary gland tumors, treated with neutrons between August 1977 and July 1984 at the Western General Hospital in Edinburgh. Patient mean age was 65 years. The most common location was the parotid gland in 18 cases. Adenoid cystic carcinoma and mucoepidermoid tumors were the most frequent histologic types. Nineteen patients suffered from an inoperable tumor. The remaining nine patients had been operated on previously, including three patients with recurrent disease and three patients with macroscopic residuals after surgery. Previous radiotherapy was carried out in two patients. All but one patient received neutron therapy alone. The target absorbed dose for most patients ranged from 13.7 to 17.8 Gy, given in 12–20 fractions. The local control rate for gross disease was 54.5%. Thirteen of the 28 patients developed distant metastases, mainly in lung. The median survival time for all patients was 19.5 months, with a range from 1 to 63 months. Severe late side effects could be observed in four patients.

Kovács et al. (1987) analyzed data from 15 patients with malignant parotid gland tumors treated with fast neutrons between 1972 and 1983 at Dresden-

Rossendorf. He compared the results of 110 patients treated with low-LET radiation between 1948 and 1981 in Budapest. The applied dose in the neutron group was 10.4 Gy on average fractionated over 10–15 days. Nine of the 15 patients were irradiated with neutrons for recurrent disease. The most common histologic types were malignant mixed tumors in five cases and squamous cell carcinomas in four cases. Local control was achieved in 13 (87%) patients compared to only 51 (46%) patients in the low-LET therapy group. Among the patients treated with neutrons, severe side effects were observed in two.

Pötter et al. (1990) published data from nine patients with adenoid cystic carcinoma irradiated since Januray 1986 at the University of Münster. Patients with residual tumors after surgery and with inoperable or recurrent malignancies were included. The first patients were treated according to a mixed-beam schedule with photons and neutrons. Later treatment was changed to definitive neutron therapy up to a total dose of 15 Gy. Local control was achieved in seven of the nine patients. The observed side effects were comparable to those from conventional photon therapy. An update of 27 patients with adenoid cystic carcinoma was reported at the EORTC 1993 Meeting in Brussels (Pötter et al. 1993). Eleven patients had lymph node or distant metastases and 24 patients had a history of previous surgery or radiotherapy or both. Twenty patients were irradiated with neutrons alone. The mean follow-up of all patients was 14 months, with a range of 2–46 months. The overall local control rate was 74%. Seven patients relapsed locally. Four patients died from disease and two patients from other causes. Acute skin reactions were mild to moderate, slightly more than experienced with photon therapy.

Prott et al. (1996) has given a final report on all the patients with salivary gland tumors which were treated at Münster until 1995. He published the data from 64 patients with adenoid cystic carcinoma and from 19 patients with other salivary gland tumors. In patients with adenoid cystic carcinoma, permanent local control was achieved in 61%. The survival probability was 76% after 2 years and 35% after 5 years. Late side effects were teleangiectasia in 67%, taste impairment in 61%, subcutaneous fibrosis in 36% and xerostomia in 32%.

Lessel (1995) from the Robert Rössle Hospital in Berlin reported results from 38 patients with salivary gland tumors of the parotid gland who had undergone neutron therapy since 1974. The most common histologic types were malignant mixed tumor, adenoid cystic carcinoma, mucoepidermoid carcinoma, acinic cell carcinoma, adenocarcinoma and squamous cell carcinoma. Eighteen patients were treated for primary tumor and 20 patients for recurrent disease. Neutron therapy was indicated in 10 patients for macroscopic residuals and in 20 patients for microscopic residuals after surgery. A total of 11 patients developed local recurrence after treatment. Distant metastases were diagnosed in 16 patients. Six patients suffered from severe late side effects (grade III or grade IV), including two patients who had been irradiated previously. The mean duration of local control was 116 months and the mean disease-free survival time was 93 months.

At the University Hospital Hamburg, 148 patients with malignant salivary gland tumors were treated between July 1977 and September 1994. One hundred patients were treated for primary tumor and 48 patients were irradiated for recurrent disease. The most common histologic type was adenoid cystic carcinoma in 114 patients. A subanalysis was done on 31 patients treated between 1977 and 1988. All patients had a history of one or more previous surgical interventions. The calculated control rate was 78% at 1 year and 65% at 2 years (Krüll et al. 1990). Based on these data, a controlled clinical trial was started in January 1991. In 1995 the results from 33 patients where published who had been treated with neutrons for recurrent salivary gland tumors (Krüll et al. 1995a). All patients had been operated on several times. Ten patients had been irradiated previously with photons or electrons. Twenty-flour patients had a macroscopic tumor mass. The median follow-up was 33 months. The survival rate for the 33 patients was 45% at 5 years. The calculated control rate for all patients was 43% at 5 years. Only in two cases were severe late side effects observed.

Another analysis was done for 74 patients with macroscopic tumor mass, who were irradiated from 1977 to 1993 (Krüll et al. 1996). Forty-seven patients were treated for an inoperable or not completely resected primary tumor and 27 patients for recurrent disease. Sixty-three patients were irradiated with neutrons alone. The remaining eight patients were treated with a mixed beam. The median dose at the reference point was 16.3 Gy, with a range from 7.2 to 19.7 Gy. The mean follow-up time for the analyzed patients was 39 months. The calculated local control rate of all patients was 54% at 3 years and 39% at 5 years. The survival rate was 66% at 3 years and 53% at 5 years. Six patients developed severe late side effects but in four patients the orbital cavity had to be included in the target volume and the resulting blindness was a pretherapeutic calculated risk.

The radiotherapists of Orléans treated 27 patients with malignant salivary gland tumors with neutrons between March 1982 and August 1993 (Favre and Breteau 1995). The most frequent histologic type was adenoid cystic carcinoma in 12 of them. Eighteen patients were irradiated for a primary tumor and nine patients for a recurrence. The indications for radiotherapy were microscopic residuals in 5 cases and macroscopic residuals in 17 cases. Four patients were irradiated in an adjuvant situation. Ten patients were treated with neutrons alone and 17 patients with neutrons and gamma-rays. A complete tumor remission was diagnosed in 14 patients, a partial remission in 10 patients and no change of tumor mass in the remaining 3 patients. The median follow-up of this collective was 22 months with a range from 1 to 78 months. At the time of analysis 11 patients were alive, 6 of them with local and/or distant failure.

At the University of Essen, 14 patients were treated with neutrons between October 1988 and February 1993 (Budach et al. 1993). The most common site of tumor was the parotid gland in eight cases. Five patients had been irradiated for a primary tumor and nine for recurrent disease. At the time of neutron therapy six patients had microscopic tumor residuals and eight patients suffered from macroscopic tumor mass. Four patients were treated with neutrons alone and eight patients with a combination of neutrons and photons. The median follow-up was 7 months. A complete remission was observed in ten patients and a partial remission in the remaining four patients. Four patients developed acute side effects classified as grade III.

From 1976 to 1993 a total of 49 patients were irradiated at the neutron facility of Heidelberg for adenoid cystic carcinomas (Engenhart et al. 1994). Thirty-one patients were treated with neutrons alone and 18 patients with a mixed beam. Twenty-three patients were irradiated for a primary tumor and 26 patients for a recurrence. Patients were treated with neutrons either postoperatively, when gross disease or extensive microscopic residuals remained, or as primary treatment without previous surgery. The minimum follow-up for the analyzed patients was 12 months. The calculated 5-year survival rates were 70% for primary tumors and 65% for recurrences. The local control rates were 75% for primary tumors and 65% for recurrent disease. The local control rates at 3 years were 70% for patients treated with neutrons alone compared to 25% for patients treated with neutrons and photons. Slight complications appeared in approximately 66% of the patients. Severe side effects could be observed in two patients and this was related to the delivered dose in one case.

In the treatment period from 1984 to August 1992 a further eight patients with salivary gland tumors but with other histologies were treated (Gademann et al. 1993). Four received neutron therapy for a primary tumor and four patients were irradiated for a recurrence. Three patients were treated with neutrons alone and the applied dose was 15.2 Gy. The median follow-up in this subgroup was 26 months. At the time of analysis six patients were still alive.

At the University of Munich, 40 patients were irradiated with neutrons (Auberger 1995), 14 male and 16 female. Different histologies were included. All were treated with combined modalities.

Discussion

The results of conventional radiotherapy are suboptimal in inoperable or not completely resected salivary gland tumors and in recurrent disease. T.W. Griffin et al. (1988) pooled the data of some international series for low-LET radiation, giving a local contral rate of 28%. He included the analyses from Fitzpatrick and Thenault (1986), Borthne et al. (1986), Rafla (1997), Fu et al. (1977), Stewart et al. (1968), Shidnia et al. (1980), Elkon et al. (1978) and Rossmann (1985). Comparable reviews were published by Wambersie (1990) and Budach (1991). Wendt et al. (1992) calculated a locoregional tumor control rate of only 22% in patients irradiated with photons after incomplete surgery.

Only a few radiotherapists, e.g., Kreitner et al. (1988) and Wang and Goodman (1991), have achieved better results. Kreitner et al. (1988) used high-dose photon therapy up to 80 Gy or interstitial irradiation with a local dose of 70 Gy. Local tumor control was observed in all treated patients.

Wang and Goodman (1991) irradiated patients with malignant salivary gland tumors with accelerated and hyperfractionated photons and applied an additional boost, including electron beam, interstitial implant, intraoral cone or submental photons. He described a local control rate of 85% at 5 years. Griffin (1991) reviewed these data, eliminated the smaller tumors from this series and counted two additional failures as local relapses. The resulting local control rate was still 71%. However, by eliminating patients, the median follow-up time was reduced to only 20 months. Longer follow-up periods are necessary for a definite conclusion to be drawn about the efficacy of accelerated photon radiation in gross salivary gland tumors.

The first data for neutron therapy in cancer were published by Stone in 1948. Missing experiences in dosimetry and biological interactions led to severe treatment complications and the use of neutrons in cancer was discouraged. Further knowledge in dosimetry and radiobiology was gained and neutron therapy restarted in Great Britain in the late 1960s.

In 1981 Battermann et al. defined the relative biological effectiveness (RBE) of different malignant tumors during the treatment of lung metastasis with neutrons or gamma-rays and measured the different growth delay. They found one of the highest RBEs in adenoid cystic carcinoma for fractionated therapy. The RBE was 8 compared to 3.0 and 3.5 for normal mesenchymal tissue, producing a therapeutic gain factor for neutrons in contrast to conventional radiotherapy of approximately 2.5. Since then clinical trials in Europe and in the United States have confirmed these radiobiological data. Only in Europe have at least 520 patients with salivary gland tumors been treated with neutrons alone or with combined modalities (Table 1). The most common histologic types are adenoid cystic carcinoma, mucoepidermoid carcinoma, malignant mixed tumor and adenocarcinoma (Table 2).

Especially in advanced salivary gland tumors, neutron therapy can improve local control. The clinical data from different centers in Europe and the United States, pooled by T.W. Griffin et al. (1988), show a local control rate of 67% for gross disease (Battermann and Mijnheer 1986; Catterall 1987; Duncan et al. 1987; Eichhorn 1981; B.R. Griffin et al. 1988; Maor et al. 1981; Ornitz et al. 1979; Saroja et al. 1987; Skolyszewski et al. 1982). A similar control rate, namely 65%, can be achieved, if only data from European series are analyzed (Table 3).

To prove the efficacy of fast neutron therapy, a randomized RTOG/MRC study was started by T.W. Griffin et al. in 1988. There was a statistically significant improvement of local control in the neutron therapy arm compared to conventional therapy and for ethical reasons this study was terminated. The final results of these study with a follow-up time of 10 years were reported by Laramore et al. in 1993 and there was still a significant benefit in local control, namely 56% for neutrons vs. 17% for photons, but there was no longer a clear difference in overall survival as previously observed. This was explained by the fact that distant metastases accounted for most of the

Table 1. Pooled number of treated patients in Europe

Reference	No. of patients
Catterall (1987)	65
Battermann and Mijnheer (1986)	32
Duncan et al. (1987)	28
Budach et al. (1993)	14
Favre and Breteau (1995)	27
Prott et al. (1996)	83
Gademann et al. (1993)	8
Krüll et al. (1997)	183
Engenhart et al. (1994)	49
Lessel 1995	38
Auberger 1995	40
Skolyszewski et al. (1982)	3
Overall	570

Table 2. Distribution of histologic types

Histology	No.	%
Adenoid cystic carcinoma	239	61.4
Mucoepidermoid carcinoma	42	10.8
Malignant mixed tumor	27	6.9
Adenocarcinoma	25	6.4
Acinic cell tumor	13	3.3
Squamous cell carcinoma	10	2.6
Others	33	8.5
Overall	389	100

Pooled data from Berlin (Lessel 1995), Edinburgh (Duncan et al. 1987), Hamburg (Krüll et al. 1995b), Heidelberg (Engenhart et al. 1994), London (Catterall 1987), Münster (Pötter et al. 1993) and Orléans (Favre and Breteau 1995).

Table 3. Pooled European data of local control in advanced salivary gland tumors

Reference	No.	Local control (%)
Catterall (1987)	65	48 (74)
Battermann and Mijnheer (1986)	32	21 (66)
Duncan et al. (1987)	22	12 (55)
Prott et al. (1996)	64	39 (61)
Kovács et al. (1987)	15	13 (87)
Krüll et al. (1995b)	74	44 (59)
Skolyszweski et al. (1982)	3	2 (67)
Overall	275	179 (65)

failures in the neutron arm. These findings are consistent with most of the published or reported uncontrolled European series. The median survival time in patients with advanced salivary gland tumors ranges between 16.5 (Duncan et al. 1987), 22 (Battermann and Mijnheer 1986), 36 (Catterall et al. 1987) and 61 (Krüll et al. 1995b) months.

The incidence of severe late side effects after neutron therapy has been high in recent publications. Based on 232 evaluable cases, Koh et al. (1989) calculated an overall complication rate for severe late side effects of 19%, with a range from 13% (Battermann and Mijnheer 1986) to 23% (Saroja et al. 1987). The main reason ist that most of the patients in the historical series had been irradiated with inadequate neutron machines. Another reason is that many patients with gross disease have a history of radical surgery or even previous radiotherapy. Finally patients with advanced tumors have a higher risk of developing side effects, because many tumors have already infiltrated critical structures. An update of different European series results in an overall complication rate of 10.7% for severe radiation-related morbidity, with a range from 4.1% (Engenhart et al. 1994) to 15.8% (Lessel 1995) (Table 4).

Modern neutron machines with integrated multileaf collimators and the use of three-dimensional treatment planning systems are now available in a few institutions and will further reduce morbidity (Stelzer et al. 1994).

With the excellent local control rates being reported, the role of surgery in gross salivary gland tumors has been newly defined. Buchholz et al. (1992) analyzed the local control rate for fast neutron therapy in previously untreated advanced salivary gland tumors. The 5-year local control rate in this subgroup was 92% compared to 51% for patients with unresectable recurrent tumors, who had undergone previous surgery one or several times. He concluded that surgery should be limited to those patients in whom the tumor can be resected with negative surgical margins.

Another argument for using neutrons especially in patients with advanced parotid gland tumors is the preservation of facial nerve function in contrast to the effect of radical surgery. Catterall (1987) observed this effect in all of the patients whose parotid gland tumors were controlled by neutrons, even in the advanced stages.

Table 4. Pooled European data of severe radiation-related morbidity

Reference	No.	%
Catterall (1987)	8/65	11.8
Battermann and Mijnheer (1986)	4/32	12.5
Duncan et al. (1987)	4/26	15.4
Engenhart et al. (1994)	2/49	4.1
Krüll et al. (1995b)	6/74	8.1
Lessel (1995)	6/38	15.8
Overall	30/284	10.6

In summary, neutron therapy offers a therapeutic gain factor compared to low-LET radiation and should be the treatment of choice in advanced salivary gland malignancies, which are inoperable or not completely resected, and in recurrent disease.

References

Armstrong JG, Harrison LB, Spiro RH, Fass DH, Strong EW, Fuks ZY (1990) Malignant tumors of major salivary gland origin: a matched pair analysis of the role of combined surgery and postoperative radiation therapy. Arch Otolaryngol Head Neck Surg 116:290–293

Auberger T (1995) EORTC Heavy-Particle Therapy Group and European Clinical Heavy-Particle Dosimetry Group (ECHED), May 3-6, Orléans

Battermann JJ, Mijnheer BJ (1986) The Amsterdam fast neutron radiotherapy project: a final report. Int J Radiat Oncol Biol Phys 12:2093–2099

Battermann JJ, Breur K, Hart G, van Peperzeel HA (1981) Observations on pulmonary metastases in patients after single doses and multiple fractions of fast neutrons and cobalt-60-gamma rays. Eur J Cancer 5:539–548

Borthne A, Kjellevold K, Kaalhus O, Vermund H (1986) Salivary gland malignant neoplasms: treatment and prognosis. Int J Radiat Oncol Biol Phys 12: 747–754

Buchholz TA, Laramore GE, Griffin BR, Koh WJ, Griffin TW (1992) The role of fast neutron radiation therapy in the management of advanced salivary gland malignant neoplasms. Cancer 69 (11):2779–2788

Budach V (1991) The role of fast neutrons in radiooncology – a critical appraisal. Strahlenther Onkol 167:677–692

Budach V, Dinges S, Stuschke M, Sack H (1993) In: Proceedings of the Workshop of EORTC Heavy-Particle Therapy Group and European Clinical Heavy-Particle Dosimetry Group (ECHED), March 11-13, Brussels

Catterall M (1987) The implications of improved treatment of malignant salivary gland tumors by fast neutron radiotherapy. Int J Radiat Oncol Biol Phys 13:1313–1318

Catterall M, Blanke RP, Rampling RP (1984) Fast neutron treatment as an alternative to radical surgery for malignant tumors of the facial area. Br Med J 289:1653–1655

Duncan WJ, Orr A, Arnott SJ, Jack JL (1987) Neutron therapy for malignant tumors of the salivary gland. A report of the Edinburgh experience. Radiother Oncol 8:97–104

Eichhorn HJ (1981) Pilot study on the applicability of neutron radiotherapy. Radiobil Radiother 3:262–292

Elkon D, Cloleman M, Hendrickson FR (1978) Radiation therapy in the treatment of malignant salivary gland tumors. Cancer 41:502–506

Engenhart R, Huber P, Debus J, Höver KH, Wannenmacher M (1994) EORTC Heavy Particle Therapy Group, Meeting at Heidelberg, Oct 21-22

Favre A, Breteau N (1995) EORTC Heavy-Particle Therapy Grouy, May 3-6, Orléans

Fitzpatrick PJ, Theriault C (1986) Malignant salivary gland tumors. Int J Radiat Oncol Biol Phys 12: 1743–1747

Fu KK, Leibel SA, Levine ML, Friedlander LM, Boles R, Phillips TL (1977) Carcinoma of the major and minor salivary glands: analysis of treatment results and sites and causes of failures. Cancer 40: 2882–2890

Gademann G, Engenhart R, Wannenmacher M (1993) Proceedings of the Workshop of EORTC Heavy-Particle Therapy Group and European Clinical Heavy-Particle Dosimetry Group (ECHED), March 11-13, Brussels

Glanzmann C (1990) Langzeit-Resultate nach postoperativer und primäerer Radiotherapie von Karzinomen der Kopfspeicheldrüsen: Bedeutung von Dosis und Zielvolumen. Strahlenther. Onkol 166: 183–189

Griffin BR, Laramore GE, Griffin TW, Eenma J (1988) Fast neutron radiotherapy for advanced malignant salivary gland tumors. Radiol Oncol 12:105–111

Griffin TW (1991) Optimal treatment for salivary gland tumors. Int J Radiat Oncol Biol Phys 3:857–858
Griffin TW, Pajak TF, Laramore GE, Duncan W, Richter MP, Hendrickson FR, Maor MH (1988) Neutrons vs photon irradiation of inoperable salivary gland tumors of an TROG-MRC cooperative randomized study. Int J Radiat Oncol Biol Phys 5: 1085–1090
Hunter RM, Davis BW, Gray GF, Rosenfeld L (1983) Primary malignant tumors of salivary gland Origin. A 52-year review. Am Surg 49:82–89
Koh WJ, Laramore G, Griffin TW, Russell K, Griffin B, Parker R, Davis L (1989) Fast neutron radiation for inoperable and recurrent salivary gland cancers. Am J Clin Ocol 12: 316–319
Kovács G, Merkle KH, Lessel A, Nemeth G, Kunde D, Vass L (1987) Ergebnisse der Bestrahlung mit verschiedenen Strahlenquellen bei der Therapie maligner Parotistumoren. Strahlenther Onkol 163:84–89
Kreitner KF, Zapf S, Collo D, Kutzner J (1988) Zum Stellenwert der Strahlentherapie in der Behandlung adenoidzystischer Karzinome im Kopf- und Halsbereich. Strahlenther Onkol 164:451–456
Krüll A, Schwarz R, Heyer D, Brockmann WP, Junker A, Schmidt R, Hübener KH (1990) Results of fast neutron therapy of adenoidcystic carcinoma of the head and neck at the neutron facility Hamburg-Eppendorf. Strahlenther Onkol 166:107–110
Krüll A, Schwarz R, Heyer D, Timm H, Schmidt R (1995a) Ergebnisse der Neutronentherapie bei rezidivierenden malignen Speicheldrüsentumoren im Kopf-Hals-Bereich. Strahlenther Onkol 5:265–271
Krüll A, Schwarz R, Engenhart R, Huber P, Lessel A, Koppe H, Favre A, Breteau N, Auberger T (1996) European results in neutron therapy of malignant salivary gland tumors. Bull Cancer/Radiother 83:125–129
Laramore GE (1987) Fast neutron radiotherpy for inoperable salivary gland tumors. Int J Radiat Oncol Biol Phys 13:1421–1423
Laramore GE, Krall JM, Griffin TW, Duncan W, Richter MP, Saroja KR, Maor MH, Davis LW (1993) Neutron versus photon irradiation for unresectable salivary gland tumors: final report of an RTOG-MRC Randomized Clinical Trial. Int J Radiat Oncol Biol Phys 2: 235–240
Lessel A (1995) EORTC Heavy-Particle Therapy Group, May 3-6, Orléans
Maor M, Hussey D, Fletcher GH, Jesse RH (1981) Fast neutron radiotherapy for locally advanced head and neck tumors. Int J Radiat Oncol Biol Phys 7:155–163
Million RR, Cassisi NJ (1994) Management of head and neck cancer. A multidisciplinary approach. Lippincott, Philadelphia, pp 711–750
North CA, Lee DJ, Piantadosi S, Zahurak M, Johns ME (1989) Carcinoma of the major salivary glands treated by surgery or surgery plus postoperative radiotherapy. Int J Radiat Oncol Biol Phys 6:1319–1326
Ornitz R, Herskovic A, Bradley E, Deyeand JA, Rogers CC (1979) Clinical observations of early and late normal tissue injury and tumor control in patients receiving fast neutron irradiation. In: Barendsen GW, Broerse J, Breur K (eds) High LET radiations in clinical radiotherapy. Pergamon New York, pp 43–50
Perez CA, Brady LW (1992) Principles and practice of radiation oncology. Lippincott, Philadelphia, pp 1–63, 657–669
Pötter R, Naszalny A, Hemprich A, Haverkamp U, Al-Dandashi C, Höver KH, Loncar I (1990) Neutronentherapie bei adenoidzystischen Karzinomen. Dtsch Z Mund Kiefer Gesichtschir 14:12–24
Pötter R, Prott FJ, Loncar I, Haverkamp U, Kovcs G, Willich N (1993) In: Proceedings of the Workshop of EORTC Heavy-Particle Therapy Group and European Clinical Heavy-Particle Dosimetry Group (ECHED), March 11-13, Brussels
Prott FJ, Haverkamp U, Willich N, Wagner W, Micke O, Pötter R (1996) Ten years of fast neutron therapy in Münster. Bull Cancer/Radiother 83:115–121
Rafla S (1977) Malignant parotid tumors: natural history and treatment. Cancer 40:136–144
Rink B, Becker D (1990) Adenoidzystisches Karzinom im Kiefer-Gesichtsbereich. Dtsch Z Mund Kiefer Gesichtschir 14:450–455

Rossmann KJ (1975) The role of radiation therapy in the treatment of parotid carcinomas. Am J Radiol 123: 492-499

Sadeghi A, Tran LM, Mark R, Sidrys J, Parker RG (1993) Minor salivary gland tumors of the head and neck: treatment strategies and prognosis. Am J Clin Oncol 16:3–8

Saroja KR, Mansell J, Hendrickson FR, Cohen L, Lennox A (1987) An update on malignant salivary gland tumors treated with neutrons at Fermilab. In J Radiat Oncol Biol Phys 13:1319–1325

Seifert G, Miehlke A, Haubrich J, Chilla R (1986) Diseases of the salivary glands. Pathology-diagnosis-treatment-facial nerve surgery. Thieme, Stuttgart

Shidnia H, Hornback HS; Hamaker R, Lingemann R (1980) Carcinoma of major salivary glands. Cancer 45:693–697

Shingaki S, Ohtake SK, Nomura T, Nakajima T (1992) The role of radiotherapy in the management of salivary gland carcinomas. J Craniomaxillofac Surg20:220–224

Skolyszewski J, Byrski E, Chrzanowska A, Gasinka A, Reinfuss M, Huczkowski J, Lazarska B, Michalowski A, Meder J (1982) A preliminary report on the clinical application of fast neutrons in Krakow. Int J Radiat Oncol Biol Phys 8:1781–1786

Spiro RH (1986) Salivary neoplasms: overview of a 35-year experience with 2,807 patients. Head Neck Surg 8:177–184

Stelzer KJ, Laramore GE, Griffin TW, Koh TW, Austin-Seymour MA, Russell KJ, Buchholz TA (1994) Fast neutron radiotherapy. Acta Oncol 33:275–280

Stewart JG, Jackson AW, Chew MC (1968) The rate of radiation therapy in the managment of malignant tumors of salivary glands. AJR AM J Roentgenol 102: 100–108

Stone RS (1948) Neutron therapy and specific ionization. AJR Am J Roentgenol 59: 77-85

Takeichi N, Hirose F, Yamamoto H (1976) Salivary gland tumors in atomic bomb survivors, Hiroshima, Japan. Cancer 38: 2462–2468

Vrielinck LJG, Ostyn F, van Damme B, van den Bogaert W, Fossion E (1988) The significance of perineural spread in adenoid cystic carcinoma of the major and minor salivary glands. Int J Oral Maxillofac Surg 17:190–193

Wambersie A (1990) Fast neutron therapy at the end of 1988 – a survey of the clinical data. Strahlenther Onkol 166:52–60

Wang CC, Goodman M (1991) Photon irradiation of unresectable carcinomas of salivary glands. Int J Radiat Oncol Biol Phys 3:569–576

Wendt TG, Rübe C, Lissner J (1992) Ergebnisse der Photonentherapie von malignen Tumoren der Ohrspeicheldrüse. Strahlenther Onkol 168:311–317

European Results of Neutron Therapy in Soft Tissue Sarcomas

R. Schwarz[1], A. Krüll[1], A. Lessel[2], R. Engenhart-Cabillic[3], A. Favre[4], F.J. Prott[5], T. Auberger[6], and R. Schmidt[1]

[1] Abteilung für Strahlentherapie und Radioonkologie, Universitätskrankenhaus Hamburg-Eppendorf, Martinistrasse 52, D-20251 Hamburg, Germany
[2] Funktionsbereich Strahlentherapie, Robert-Rössle-Klinik am Max-Delbrück-Centrum Berlin-Buch, Medizinische Fakultät der Humboldt-Universität zu Berlin, Lindenberger Weg 80, D-13125 Berlin-Buch, Germany
[3] Abteilung für Strahlentherapie, Medizinisches Zentrum für Radiologie, Baldingerstrasse, D-35033 Münster, Germany
[4] Service de Radiothérapie, Centre Hospitalier Régional d'Orléans, Hôpital de la Source, BP 6709, F-45067 Orléans cedex 2, France
[5] Klinik und Poliklinik für Strahlentherapie und Radioonkologie, Westfälische Wilhelms-Universität, Albert-Schweitzer-Strasse 33, D-48129 Münster, Germany
[6] Klinik und Poliklinik für Strahlentherapie und Radiologische Onkologie, Klinikum Rechts der Isar der Technischen Universität München, Ismaninger Strasse 22, D-81675 Munich, Germany

Abstract

The purpose of this publication is to review the data of European neutron therapy facilities in the treatment of soft tissue sarcoma. In the past locally advanced soft tissue sarcoma was a common indication for neutron therapy in several European centers. Data from 1171 patients treated at 11 centers were analysed on the basis of published data or personal communication. Neutron therapy was of benefit especially for patients with primary radiotherapy of inoperable tumors, and with macroscopic disease after surgery. Local control was about 50% in these patients. These results are superior to those from photon therapy from the literature. Results for postoperative neutron therapy after R0- and R1-resection are similiar to those from photon therapy. The incidence of late side effects was considerably high in some neutron series. This is influenced by the treatment of most patients with first-generation machines of limited technical standard, and that in most cases large treatment volumes were irradiated.

Introduction

Soft tissue sarcomas constitute only 1% of all malignant tumors. They consist of a considerable number of histologic subtypes and are designated in most instances according to the pattern of differentiation and presumably

Recent Results in Cancer Research, Vol. 150

the tissue of origin, i.e., fibrosarcoma, liposarcoma, leiomyosarcoma, synovial sarcoma, malignant fibrous histiocytoma, etc. The clinical courses of these vary and so do their responses to treatment, whether by surgery, radiation, or drugs. The natural history and response of soft tissue sarcomas to the various therapies are much less defined than for many other tumors.

Well-operable soft tissue sarcomas are treated by radical resection. Local resection with a varying safety margin is an inadequate surgical procedure, leading to local recurrences in the range of 55–90% (Cantin et al. 1968; Markhede et al. 1982; Shin et al. 1975; Simon and Enneking 1976). For more advanced lesions the treatment will often be a combination of doubtful radical or nonradical resection and preoperative or postoperative radiotherapy, in some cases supplemented by chemotherapy. Postoperative high-dose photon irradiation in the range of 60–66 Gy at 6–7.5 weeks after wide excision of the tumor improves the local control rates to 72% n 90% for completely resected tumors (Lindberg et al. 1981; Suit 1983; Suit and Russell 1975). A special problem is that of achieving local control of inoperable tumors or gross residual tumor after surgery. These tumors may often grow large before discovery by the patient. Soft tissue sarcomas are considered radioresistant because of the poor outcome of only 0–56% local control 5 years after definite high-dose photon irradiation (Duncan and Dewar 1986; Leibel et al. 1982; McNeer et al. 1968; Tepper and Suit 1985; Windeyer et al. 1966). It is generally believed that the radioresistance of soft tissue sarcomas ist associated with the large capacity of the tumor cells to accumulate and repair damage. Existence of a significant hypoxic fraction within the tumor cell population would further enhance this resistance to conventional radiotherapy. The suboptimal results have led to trials with high linear energy transfer (LET) forms of radiation, such as fast neutron radiation therapy.

Radiobiology

Neutrons are subatomic particles that interact biologically with tissue differently than conventional irradiation. The first data presented on the use of neutron irradiation in the treatment of cancer were published in 1942 (Stone and Larkin 1942). The limited understanding of the neutron-matter interaction and neutron dosimetry led to severe treatment complications, and the use of neutrons for irradiation of cancer was discouraged. Our knowledge of the dosimetry and radiobiology of neutrons has improved since then and neutron therapy was restarted in Great Britain in the late 1960s at Hammersmith Hospital.

Fast neutrons have different biological effects from X-rays and these differences result in therapeutic advantages in treating locally advanced or "radioresistant" tumor at certain sites. The reasons for these high relative biological effectiveness (RBE) values are manifold. Neutrons interact with the atomic nuclei of cells and produce densely ionizing protons. These protons transfer high amounts of energy to the tumor and inflict a significant percentage of single-hit double-stranded DNA damage. This type of radiation-induced damage generally is considered lethal and unreparable. Another favorable consequence of the higher percentage of unreparable double-stranded DNA damage that neutrons cause is a reduced variability of inflicting a lethal event across the different phases of the cell cycle. Other important radiation biological concepts are the oxygen enhancement ratio and tumor hypoxia. Photon radiation causes most of its cellular damage through the generation of free radicals. This process depends biochemically on the presence of oxygen in the tumor. Neutrons depend less on oxygen to act as a mediator and have a greater tendency to cause death through a direct interaction with the tumor cells.

Overview

During the last 25 years a large number of patients with soft tissue sarcomas have been treated with fast neutrons worldwide. The purpose of this publication is to review the data of European neutron therapy facilities in the treatment of soft tissue sarcomas. All European neutron centers have treated patients with soft tissue sarcomas.

Data from 11 European neutron therapy centers could be obtained, either from publications or by personal communication in preparation for the EORTC '95 – Heavy Particle Therapy Group Meeting at Orléans. Publications suitable for this review were from the neutron therapy centers at Amsterdam (Battermann and Breur 1981), Edinburgh (Duncan et al. 1986), Essen (Budach et al. 1990; Schmitt et al. 1989), Hamburg (Franke 1987), Hammersmith (Pickering et al. 1987), and Louvain-la Neuve (Richard et al. 1985). Personal communications and analyses suitable for use were from Berlin (Lessel 1995, pers. comm.), Hamburg (Schwarz and Krüll, unpubl. results), Heidelberg (Engenhart 1995, pers. comm., Auberger 1995, pers. comm.); Münster (Prott 1995, pers. comm.), and Orléans (Favre 1995, pers. comm.). Some neutron facilities have closed during recent years, while others are continuing their work.

Table 1 shows the treatment center, the reporting period and the number of treated patients. The largest groups of patients with soft tissue sarcomas were treated at Hamburg (319 patients), Essen (261 patients), and Berlin-Buch (221 patients). Lists from different overviews (Schmidt et al. 1993; Wambersie and Richard 1991) demonstrate that soft tissue sarcoma has been the most frequent tumor entity treated with neutrons during the last 25 years in Europe. In spite of this fact here has been only one attempt to carry out a randomized study comparing neutrons and photons for postoperative irradiation in soft tissue sarcoma (Glaholm and Harmer 1988). Fourteen patients were entered into this prospective trial in Great Britain, but it was decided at an early stage to terminate the trial because of unacceptable late tissue damage.

Table 1. Neutron therapy in soft tissue sarcomas. Institution, period of treatment, and number of treated patients

Institution	Period	No. of patients
Amsterdam	1975–1981	22
Berlin	1975–1994	221
Edinburgh	1977–1982	16
Essen I	1978–1983	221
Essen II	1983–1988	40
Hamburg I	1976–1980	86
Hamburg II	1984–1995	233
Hammersmith	1971–1983	66
Heidelberg	1975–1994	107
Louvain-la-Neuve	1979–1983	46
Münster	1986–1994	42
Munich		26
Orléans	1982–1993	45

Patient characteristics

Complete data concerning the age of the patients were available from four centers. The median age of the treated patients was 44.4 years at Berlin-Buch (Lessel 1995, pers. comm.), 57.2 years at Essen (Budach et al. 1990), 50.6 years at Hamburg (Schwarz and Krüll, unpubl. results), 54 years at Hammersmith (Pickering et al. 1987), and 46 years at Orléans (Favre 1995, pers. comm.). The ranges were 18–87 years at Berlin-Buch (Lessel 1995, pers. comm.), 23–72 years (Schmitt et al. 1989), and 19–84 years (Budach et al. 1990) at Essen, 13–86 years at Hamburg (Schwarz and Krüll, unpubl. results), 10–89 years at Hammersmith (Pickering at al. 1987), and 15–78 years at Orléans (Favre 1995, pers. comm.).

Location

Because of the ubiquitous nature of the connective tissues, soft tissue sarcomas arise anywhere in the body. A review of patients with soft tissue sarcomas from the literature (Abbas et al. 1981; Ferrell and Frable 1972; Hare and Cerny 1963; Lindberg et al. 1981; Potter et al. 1985; Sears et al. 1980; Shieber and Graham 1962; Torosian et al. 1988) demonstrates that soft tissue tumors are mainly located in the lower and upper extremities. Fourteen percent of the tumors were located in the upper extremity, 45% in the lower extremity, 32% at the trunk or the retroperitoneum, and 9% in the head and neck. Exact data concerning the neutron groups were available from seven centers. The percentage of extremity tumors was 36% at Amsteram (Battermann and Breur 1981), 64% at Berlin-Buch (Lessel 1995, pers. comm.), 67.5% at Essen

(Budach et al. 1990), 62% at Hamburg (Schwarz and Krüll, unpubl. results), 65% at Hammersmith (Pickering et al. 1987), 56% at Louvain-la-Neuve, excluding retroperitoneal tumors (Richard et al. 1985), and 42% at Orléans (Favre 1995, pers. comm.). This distribution is comparable to the above-mentioned data.

Histology and Grading

Soft tissue sarcomas consist of a considerable number of different histologic subtypes. A pathologic classification of soft tissue tumors, based on the cell origin of each tumor, was suggested by Enzinger and Weiss (1988). The incidence of different histologies was available from seven neutron therapy facilities (Table 2). In a wide variety of different histologies the predominance of some histologies such as liposarcoma, fibrosarcoma, malignant fibrous histiocytoma, malignant schwannoma, synovial sarcoma, and leiomyosarcoma seems to be comparable to non-neutron therapy data (Collins et al. 1986; Lindberg et al. 1981; Potter et al. 1985; Russel et al. 1975; Shin et al. 1975; Simon and Enneking 1976; Suit 1983).

The grading of the tumor ist a well-estabilshed factor influencing the outcome of the disease especially to probability of distant metastasis. Different histologic features for the grading are known (Alvegard et al. 1989; Coindre et al. 1986; Costa et al. 1984; Lack et al. 1989; Trojani et al. 1984). Mostly used criteria for grading are number of mitoses and degree of necrosis.

Data for grading were available from Essen (Budach et al. 1990; Schmitt et al. 1989), Hamburg (Schwarz and Krüll, unpubl. results), and Hammersmith (Pickering et al. 1987). The probability from of grade 1 was 25% and 30%

Table 2. Histology of soft tissue sarcomas

Institution	Liposarcoma	Fibrosarcoma	Malignant fibrous histiocytoma	Leiomyosarcoma	Malignant schwannoma	Chondrosarcoma	Others
Berlin	58	44	29	12			78
Edinburgh	3	5		6			2
Essen I	54	34	21	16			96
Essen II	14	4	12				10
Hamburg I	16	22	7		6		35
Hamburg II	67	29	44	30	19		44
Hammersmith	16	4	14	6	6		20
Heidelberg	41	15	13	6		14	18
Louvain	10	9	5			4	18
Orléans	5	8	10	4			18
Total	367	174	155	80	31	18	339
Percentage	31.5%	14.9%	13.4%	6.9%	2.7%	1.5%	29.1%

Table 3. Tumor stage "T" for soft tissue sarcomas

Institution	T1	T2	T3
Berlin	69	129	
Essen I	15	100	106
Essen II	7	33	
Hamburg II	50	183	
Louvain	8	16	17
Total	149	461	123

(Essen, Budach et al. 1990), 29% (Hamburg, Schwarz and Krüll, unpubl. results), and 40% (Hammersmith, Pickerung et al. 1987); of grade 2 48% and 230% (Essen, Budach et al. 1990), 34% (Hamburg, Schwarz and Krüll, unpubl. results), and 24% (Hammersmith, Pickering et al. 1987); and of grade 3 27% and 40% (Essen, Budach et al. 1990), 37% (Hamburg, Schwarz and Krüll, unpubl. results), and 36% (Hammersmith, Pickering et al. 1987).

Tumor size

Locally advanced soft tissue tumors were predominantly represented in the neutron groups. Data are shown in Table 3 (Budach et al. 1990; Richard et al. 1985; Schmitt et al. 1989; Schwarz and Krüll, unpubl. results; Lessel 1995, pers. comm.). All authors agreed that especially in advanced lesions (for example T2 lesions, or the T3 lesion from the old classification) neutron therapy is indicated.

Residual tumor

Table 4 shows that a high percentage of patients had macroscopic tumor. Of 932 evaluable patients, 429 patients (46%) hat gross tumor at the time of radiotherapy. These patients had inoperable primary or recurrent tumor and were treated primarily by neutron irradiation, or they had macroscopic (tumors) residuals (R2) after surgery and were irradiated postoperatively. Moreover, 278 patients (29.8%) were irradiated postoperatively for microscopic disease after surgery and 225 patients (24.2%) were referred after R0 resection of large tumors. Although detailed information ist not given in some papers, it can be assumed that the majority of patients were operated on before neutron therapy was given. A minority of patients had been previously irradiated with photons or electrons for a longer time before neutron therapy: at Berlin, 4/221 patients (Lessel 1995, pers. comm.); at Hamburg, 14/233 patients (Schwarz and Krüll, unpubl. results); and at Hammersmith, 10/66 patients (Pickering et al. 1987). Chemotherapy was not a definite component of the multimodal therapy, but some of the neutron patients received chemotherapy, especially as palliative

Table 4. Tumor mass: residual tumor classification

Institution	R0	R1	R2
Berlin	93	20	96
Edinburgh		4	12
Essen I	23	104	94
Essen II		31	9
Hamburg I		4	17
Hamburg II	95	49	82
Hammersmith		16	50
Louvain		27	19
Münster	9	12	21
Orléans	5	11	29
Total	225	278	429
Percentage	24.2%	29.8%	46.0%

treatment for distant metastases, for example, 41/233 patients received chemotherapy at Hamburg (Pickering et al. 1987).

Radiotherapeutic Modalities

Two different treatment modalities were used: (a) neutron therapy and (b) neutron therapy as boost and photon therapy. At Essen, patients with macroscopic tumor were irradiated with neutrons with doses of 16 Gy. After R0- or R1-resection photons with doses of 40 Gy were used in combination with a neutron boost of 4–6 Gy (Budach et al. 1990; Schmitt et al. 1989). At Hamburg (Schwarz and Krüll, unpubl. results), neutrons were preferred; 178/233 patients were treated with neutrons (35 of them with doses less than 14.0 Gy, 64 patients with doses $\geqslant$ 14.0 Gy and < 16.0 Gy, and 79 patients with doses $\geqslant$ 16.0 Gy). Fifty-five patients were irradiated with photons for the large volume and a neutron boost with different doses. At Hammersmith pure neutrons with a dose of 15.6 Gy and 3×1.3 Gy/week were preferred. At Berlin patients were treated with neutrons using two different schedules with 10×1.2 Gy or 5×2.4 Gy in a split course in comibnation with photons with a minimum dose of 20 Gy. The total dose for neutron therapy varied and was dependent on the location of the target volume and the individual situtation of the patient. At Edinburgh neutrons with doses in a range of 13.8–18.3 Gy were used. At Louvain-la-Neuve patients were treated with neutrons with doses of 18.7 or 23.2 Gy. At Amsteram neutrons with a minimum of 19 Gy or a neutron/photon combination were used. At Münster patients were treated with neutrons or photons. At Orléan neutrons plus photons were preferred in 39/45 patients with total doses of 45-70 Gy eq, with a mean dose of 61 Gy eq. Six patients were treated with neutrons with total doses of about 17 Gy.

Follow-up

Mean follow-up times were indicated in a few analyses: Essen (Budach et al. 1990) 29.3 months, Essen "long" follow-up (Schmitt et al. 1989), Hamburg 45 months (Schwarz and Krüll, unpubl. results), Hammersmith 24 months (Pickering et al. 1987), Louvain-la-Neuve 25.9 months (Richard et al. 1985), and Münster 20 months (Prott 1995, pers. comm.).

Local Control

Data were available for nine tratment facilities. Local control depends on different variables. At Essen (Schmitt et al. 1989) local control rates for R0 situation were 87%, for R1 situation 78% and for patients with gross tumor 56%. These results were confirmed for Essen (Budach et al. 1990) with 3-year local control rates of 80% in R1 situation and 60% in R2 situation. Similar results were found at Hamburg (Schwarz and Krüll, unpubl. results) with 5-year rates of 71.9% for the whole group, 87.7% for adjuvant radiotherapy, 64.9% for radiotherapy for microscopic disease, and 50.6% for neutron therapy for gross tumor. At Louvain-la-Neuve, Richard (Richard et al. 1985) found local control rates of 93% for R0/R1 situation, but only 21% in patients irradiated for gross tumor. At Berlin-Buch (Lessel 1995, pers. comm.) the local control rate was 66% for the whole group rates were 75% for R0/R1-situation, and 41.7% for gross tumor. At Hammersmith (Pickering et al. 1987) the local control rates were 94% after 5 years for R0/R1 situations and 52% for R2-situations. At Münster (Prott 1995, pers. comm.) 4 of 38 patients developed local recurrences and at Amsterdam only in 8 of 22 patients was local control achieved. At Orléans (Favre 1995, pers. comm.) local control was possible in 27 of 45 patients (45%). Local control rates decrease with tumor size and local tumor stage. Pickering et al. (1987) described local control rates of 71% for tumors with a diameter of less than 5 cm, 66% for 5-10 cm, 36% for 10-15 cm, and 41% for 15-25 cm. Schmitt published local control rates of 93% in T1 tumors, 82% in T2 tumors, and 64% in T3 tumors. For tumors with a diameter of 5-10 cm Schmitt et al. (1989) described 76% local control and for patients with tumors larger than 10 cm in diameter 42% local control. Richard et al. (1985) reported local control for 3 of 4 patients with T1 tumors, but in only 1 of 11 patients with T3 tumor and gross tumor was local control achieved. At Hamburg (Schwarz and Krüll, unpubl. results) the control rates for T1 tumors were 90% after 5 years and 65.9% for T2 tumors.

Grading had an influence on local control. The 3-year local control rates were 83% for G1, 67% for G2, and 59% for G3/4 at Essen (Budach et al. 1990). At Hammersmith (Pickering et al. 1987) the local control rates were 66% in G1 tumors with gross tumor vs. 54% in G2 vs. 38% in G3. At Hamburg (Schwarz and Krüll, unpubl. results) local control rates were similar at 89.9% after 5 years for G1, 74.3% for G2, and 51.4% for G3.

Patients receiving neutron therapy for primary tumors had a better prognosis than patients treated for local recurrences: At Hamburg the local control rate was 77.9% for primary tumors vs. 59.5% for recurrent tumors, a difference which was statistically significant.

Survival

Survival rates show differences from center to center, which may be a consequence of different probabilities of risk factors and a consequence of the different treatment seleciton. Different methods were used to calculate the survival rates, which were as follows: Berlin-Buch (Lessel 1995, pers. comm.), total survival after 5 years about 77%; Essen (Budach et al. 1990), 91% overall survival after 3 years; Hamburg (Schwarz and Krüll, unpubl. results) 61.4% actuarial survival after 5 years; Hammersmith (Pickering et al. 1987) 30% of all patients lived at the time of analysis; Louvain-la-Neuve (Richard et al. 1985) 72%; Münster (Prott 1905, pers. comm) 53% after 5 years; and at Orléans (Favre 1995, pers. comm.) 18 of 45 patients were alive at the time of analysis. The survival depends on factors such as grading, residual tumor classification, tumor size, and primary or recurrent tumor as an indication for neutron therapy. The 5-year disease-free survival rates at Essen (Schmitt et al. 1989) for grade 1 tumors were 74%, for grade 2 tumors 48%, and for grade 3 tumors 22%. At Hamburg (Schwarz and Krüll, unpubl. results) actuarial 5-year survival rates were 83.2% for grade 1, 76.8% for grade 2, and 32.8% for grade 3 tumors. Data for the Münster group (Prott 1995, pers. comm.) were 71% for grade 1 and 68% for grade 2 tumors. Three of four patients with grade 3 tumors died.

Results according to residual tumor classification at Essen (31%, disease free survival after 5 years) were 87% for R0, 65% for R1, and 26% for R2. The figures from Hamburg were comparable (Schwarz and Krüll, unpubl. results): R0 79.7%, R1 72.7%, and R2 34.4%. At Louvain-la-Neuve (Richard et al. 1985) the survival was 74% for R0/R1. For the Münster group (Prott 1995, pers. comm.) the results were 73% at 5 years in R0/R1 and 33% in R2.

Tumor stages influence survival as can be shown by the results from the Essen center (Schmitt et al. 1989) where disease-free survival at 5 years was 60% for T1, 71% for T2 and 29% for T3. Actuarial survival rates at 5 years from Hamburg (Schwarz and Krüll, unpubl. results) were 78.9% for T1 and 56.3% for T2. Patients treated with fast neutrons for primary tumors at Hamburg (Schwarz and Krüll, unpubl. results) had 5-year survival rates of 69.4% vs. 48% for patients treated for local recurrence.

Metastasis

At Berlin-Buch (Lessel 1995, pers. comm.) distant metastases developed in 25% of patients (56/221). Sixty-five patients (27.8%) developed distant metastases at Hamburg (Schwarz and Krüll, unpubl. results). Pickering et al. (1987)

reported that 33 of 66 patients died of metastases at Hammersmith. Prott (1995, pers. comm.) reported a 5-year probability of 31% for distant metastases. Data from Essen (Schmitt et al. 1989) and Hamburg (Schwarz and Krüll, unpubl. results) demonstrate that the incidence of metastases depends on the tumor grade. The incidence was 7% (Essen) and 7.8% (Hamburg) in G1 tumors, 21% and 28% in G2 tumors, and 53% and 47.5% in G3 tumors. The risk of metastasis was higher for patients with gross tumor at the beginning of neutron therapy (Hamburg, 37% for R2 situation) than for patients with R0 or R1 situation (24% and 20.4%). The incidence of metastatic disease correlated with increasing stage, for example at Essen (Schmitt et al. 1989) metastases developed in 27% of T1, 19% of T2, and 43% of T3 tumors.

Radiation Morbidity

Early and late radiation morbidity was reported differently. Lessel (1995, pers. comm.) reported for Berlin-Buch that acute toxicity was tolerable. Using the WHO classification no grade 3 or grade 4 toxicity occurred. The majority of reactions concerned the skin and were grade 1 reactions. At Hamburg (Schwarz and Krüll, unpubl. results) 157/233 patients developed grade 1 (RTOG classification), 79/233 developed grade 2, 3 patients grade 3, and 2 patients grade 4 acute side effects. In the other reports data on acute side effects were not given.

Battermann and Breur's (1981) experiences with late side effects were discouraging, with 6 of 22 patients developing severe side effects including skin and bowel damage. Lessel (1995, pers. comm.) reported a higher rate of complications mainly at the skin and subcutaneous tissue level. Two years after radiotherapy 13% of the patients had slight fibrosis, 26% had moderate fibrosis, and 40% had severe fibrosis. Incidence of radiation-related ulcerations was 26%. The probability of osteoradionecrosis was 0.6%. A few other complications occurred in different organs such as lung or bowel. Colleagues from Berlin-Buch noted that the risk of skin ulceration was correlated with a higher skin dose of neutrons up to 20 Gy.

Duncan et al. (1986) assessed late radiation morbidity according to the RTOG/EORTC scoring system for 18 of his patients, including patients treated for bone sarcomas. Eight of these 18 patients developed severe late skin reactions and 7 also showed severe induration of the subcutaneous tissue. One patient developed mucosa necrosis following neutron therapy for a fibrosacroma of the tonsil. A total of 50% of the Edinburgh patients had severe late radiation morbidity. Duncan saw a trend towards higher complication rates for higer dosage levels and for larger treatment volumes.

Schmitt et al. (1989) reported an incidence of severe late radiation effects scored as grades 3 or 4 (RTOG/EORTC scoring system) of 7% for neutron-boost irradiation and of 28% for neutron irradiation at Essen. This is reflected to some extent in the group of poor functional results. In a second group from Essen (Budach et al. 1990) the rate of severe late side effects was again higher for patients receiving neutron therapy versus neutron boost: 29% vs. 18%. The authors from Essen conclude that neutron boost irradiation should be preferred to avoid high complication rates.

At Hamburg (Schwarz and Krüll, unpubl. results) the RTOG/EORTC system was used for scoring of late side effects. The total incidence of grad 3, 4, or 5 late effects was 9.4% of all patients. this includes 7 grade 3 reactions, 14 grade 4 reactions, and 1 grade 5 reaction in a patient treated for a perianal soft tissue sarcoma. Grade 1 late effects were listed in 111 cases, and grade 2 late effects in 95 cases. The incidence of grade 3, 4 or 5 late effects was 8.4% in the group with neutron therapy and 12.7% in the group with neutron and photon therapy. The risk of developing severe late reactions was correlated with large treament volumes and dose inhomogeneity.

At Hammersmith (Pickering et al. 1987) late effects were scored using a tree-level system. Level II is equivalent to RTOG/EORTC grade 3/4 and level III to RTOG/EORTC grade 4. Twenty-seven patients developed late complications of level II and level III related to neutron therapy. Complications were related to the surface area of treated skin. The complication rate was 13%, for less than 200 cm^2, rising to 59% for larger areas. Pretreatment such as multiple operations, previous X-ray therapy, and hot spots in the dose distribution were further important factors.

Richard et al. (1985, Louvain-la-Neuve) observed severe complications in ten cases (22%). The risk of severe late effects was correlated to field size.

At Münster (Prott 1995, pers. comm.) no severe late effects were observed. At Orléans the incidence of severe late effects was 6.6%.

Conclusions

Soft tissue sarcomas were treated by all the European centers. The sarcomas are a particularly interesting group of tumors in that they span a spectrum from very slowly growing localized lesions to highly aggressive rapidly growing neoplasms with a high rate of distant metastases. As shown above, neutron therapy might offer advantages over conventional radiotherapy. So far, there have been no randomized trials demonstrating a benefit of neutrons over conventional high-dosage photon therapy. This review of European results for neutron therapy of soft tissue sarcomas demonstrates its advantage especially for patients with macroscopic tumor and for patients receiving primary radiotherapy of inoperable tumors. Macroscopic residuals after inadequate surgery as well as comparions with historical controls have also indicated a significant advantage. As shown above the local control rates were about 50%. especially in the large series from Berlin, Essen and Hamburg. These encouraging results are similar to those of several competing experimental modalities such as radiochemotherapy, hyperthermia, or intraoperative radiotherapy. One must accept that not only patients with macroscopic tumors will have been irradiated with neutrons, but also many patients in the adjuvant situation (R0) or for microscopic residuals have been treated with fast neutrons in Europe.

The 87% local control rates for R0 situation from Essen (Schmitt et al. 1989) and Hamburg (Schwarz and Krüll, unpubl. results) are excellent and are comparable to the best photon results in postoperative radiotherapy. The treatment results reflect the great variation in soft tissue sarcomas with patient age, histology, grading, tumor location, tumor size, tumor stage, and pretreatment. This great variation of important factors in a rare tumor is the main reason a randomized trial has not been carried out. Furthermore in the past many of the treatment machines were first-generation machines of a lower technical standard. This has influenced the high rates of severe late effects, which vary between 6.6% and 50%. In the case of neutron therapy for gross tumors, higher complication rates might be tolerable in relation to the alternative of primary amputation for many extremity sarcomas and local control rates of about 50% after neutron therapy. For adjuvant radiotherapy a higher complication rate than in photon therapy might not be tolerable. Further factors affecting late complications are pretreatment status, large treatment volumes, and dose inhomogeneities. At Essen a higher incidence of severe late effects was observed in the neutron group, at 28%, vs. 7% after the neutron boost. This difference was not confirmed by the Hamburg data, which showed incidences of severe late effects of 8.4% and 12.7% for neutron therapy and neutron boost therapy. Total doses cannot be compared easily from center to center. Some authors have shown for gross tumors that higer dose levels improve local control, but with the consequence of increasing side effects. Especially in this tumor type the treatment results are affected by many factors other than radiotherapeutic factors.

References

Abbas JS, Holyoke ED, Moore R et al. (1981) The surgical treatment and outcome of soft tissue sarcoma. Arch Surg 116:765–769

Alvegard TA, Berg NO (1989) Histopathological peer review of high-grade soft tissue sarcoma: the Scandinavian Sarcoma Group experience. J Clin Oncol 7:1845–1851

Battermann JJ, Breuer K (1981) Fast neutron radiotherapy for locally advanced sarcomas. Int J Radiat Oncol Biol Phys 7:1051–1053

Budach V, Dinges S, Bamberg M, Baumhoer W, Donhuijsen K, Sack H (1990) Neutron boost irradiation of soft tissue and chondrosarcomas at the West German Tumour Centre in Essen. Strahlenther Onkol 166:63–68

Cantin J, McNeer GP, Chu FC, Booher RJ (1968) The problem of local recurrence after treatment of soft tissue sarcoma. Ann Surg 168:47–53

Coindre JM, Trojani M, Contesso G et al. (1986) Reproducibility of a histopathologic grading system of adult soft tissue sarcoma. Cancer 58:306–309

Collins C, Hajdu SI, Godbold J et al. (1986) Localized, operable soft tissue sarcoma of the lower extremity. Arch Surg 121:1425–1433

Costa J, Wesley RA, Glatstein E et al. (1984) The grading of soft tissue sarcomas. Results of a clinicohistopathologic correlation in a series of 163 cases. Cancer 53:530–541

Duncan W, Dewar JA (1986) A retrospective study of the role of radiotherapy in the treatment of soft tissue sarcoma. Clin Radiol 37:629–632

Duncan W, Arnott SJ, Jack WJL (1986) The Edinburgh experience of treating sarcomas of soft tissues and bone with neutron irradiation. Clin Radiol 36:317–320

Enzinger FM, Weiss SW (1988) Soft tissue tumors. Mosby, St Lous

Ferrell HW, Frable WJ (1972) Soft part sarcomas revisited. Review and comparison of a second series. Cancer 30:475–480
Franke HD (1987) Clinical experiences with treatment of more than 500 patients with fast neutrons (DT, 14 MeV) since 1976 in Hamburg-Eppendorf. In: Lapis K, Eckhart S (eds) Lectures and symposia. 14th International Cancer Congress, Budapest 1986. Karger, Basel, pp 93–106
Glaholm J, Harmer C (1988) Soft-tissue sarcoma: neutrons versus photons for postoperative irradiation. Br J Radiol 61:829–834
Hare HF, Cerny MF (1963) Soft tissue sarcoma: a review of 200 cases. Cancer 16:1332
Lack EE, Steinberg SM White DE et al. (1989) Extremity soft tissue sarcomas: analysis of prognostic variables in 300 cases and evaluation of tumor necrosis as a factor in stratifying higher-grade sarcomas. J Surg Oncol 41:263–273
Leibel SA; Tranbaugh RF, Beckstead JH, Bovill EG, Phillips TL (1982) Soft tissue sarcomas of the extremities. Cancer 50:1076–1083
Lindberg RD, Martin RG, Romsdahl MM, Barkley HT (1981) Conservative surgery and postoperative radiotherapy in 300 adults with soft tissue sarcomas. Cancer 47:2391–2397
Markhede G, Angervall L, Stener B (1982) A multivariate analysis of the prognosis after surgical treatment of malignant soft tissue tumors. Cancer 49:1721–1733
McNeer GP, Cantin J, Chu F, Nickson JJ (1968) Effectiveness of radiation therapy in the management of sarcoma of the soft somatic tissues. Cancer 22:391–397
Pickering DG, Stewart JS, Rampling DG, Errington RD, Stamp G, Chia V (1987) Fast neutron therapy for soft tissue sarcoma. Int J Radiat Oncol Biol Phys 13:1489–1495
Potter DA, Glenn J, Kinsella T et al. (1985) Patterns of recurrence in patients with high-grade soft tissue sarcomas. J Clin Oncol 3:353–366
Richard F, Renard L, Wambersie A (1985) Results of neutron therapy of locally advanced soft tissue sarcomas at Louvain-la-Neuve. Strahlenther Onkol 161:794–796
Russell WO, Cohen J, Enziner F et al. (1977) A clinical and pathological staging system for soft tissue sarcomas. Cancer 40:1562–1570
Schmidt R, Rassow J, Haverkamp U et al. (1993) Physikalisch-technische Perspektive der Neutronentherapie in Deutschland. Strahlenther Onkol 169:171–178
Schmitt G, Mills EED, Levin V, Pape H, Smit BJ, Zamboglu N (1989) the role of neutrons in the treatment of soft tissue sarcomas. Cancer 64:2064–2068
Sears HF, Hopson R, Inouye W et al. (1980) Analysis of staging and management of patients with sarcoma. Ann Surg 191:488–493
Shieber W, Graham P (1962) An experience with sarcomas of the soft tissues in adults. Surgery 52:295
Shiu MH, Castro EB, Hajdu SI, Fortner JG (1975) Surgical treatment of 297 soft tissue sarcomas of the lower extremity. Ann Surg 11: 597–602
Simon MA, Enneking WF (1976) The management of soft tissue sarcomas of the extremities. J Bone Joint Surg [Am] 58:317
Stone RS, Larkin JC Jr (1942) Treatment of cancer with fast neutrons. Radiology 39:608–620
Suit HD (1983) Patterns of failure after treatment of soft tissue sarcomas by radical surgery or by conservative surgery and radiation. Cancer Treat Symp 2:241–246
Suit HD, Russell WO (1975) Radiation therapy of soft tissue sarcomas. Cancer 36:759–764
Tepper JE, Suit HD (1985) Radiation therapy alone for sarcoma of soft tissue. Cancer 56:475–479
Torosian MH, Friedrich C, Godbold J et al. (1988) Initial characteristics and prognostic factors in patients with and without metastatic disease. Semin Surg Oncol 4:13–19
Trojani M, Contesso G, Coindre JM et al (1984) Soft tissue sarcomas of adults. Study of pathological prognostic variables and definition of a histophatological grading system. Int J Cancer 33:37–42
Wambersie A, Richard F (1991) Present status of fast neutron therapy. Survey of the clinical data and of the clinical research programmes. Coordinated Research Programme "Nuclear data needed for neutron therapy", IAEA, Vienna
Windeyer W, Dische S, Mansfield CM (1966) The place of radiotherapy in the management of fibrosarcoma of the soft tissues. Clin Radiol 17:32–40

Use of Neutron Therapy in the Management of Locally Advanced Nonresectable Primary or Recurrent Rectal Cancer

R. Engenhart-Cabillic[1], J. Debus[2], F.J. Prott[3], R. Pötter[3], K.H. Höver[2], N. Breteau[4], and A. Krüll[5]

[1] Department of Radiotherapy, Medical Radiology Center, Baldingerstrasse, D-35033 Marburg, Germany
[2] Department of Radiation Oncology, German Cancer Research Center, Im Neuenheimer Feld 280, D-69120 Heidelberg, Germany
[3] Department of Radio-oncology, University of Münster, Albert-Schweitzer-Strasse 33, D-48129 Münster, Germany
[4] Centre Hospitalier Régional Orléans, Cyclotron du CNRS Orléans, BP 67009, F-45067 Orléans cedex 2, France
[5] Department of Radiooncology, University of Hamburg, Martinistrasse 52, D-20246 Hamburg, Germany

Abstract

Inoperable locally advanced or inoperable recurrent rectal cancer is a difficult problem. Tenesmus, discharge, bleeding and pelvic pain are frequently present and often are associated with infiltration of the sacral plexus. The value of radiotherapy in managing such patients is being appreciated, although up to 40% of the treated patients have no symptomatic response. Improvement in tumor response and control has been scored through efforts to overcome the radioresistance of the hypoxic tumor cells by neutron irradiation. This article is an account of the activity of neutron radiotherapy in such patients. Over 350 patients were entered in studies comparing neutrons used alone and neutrons used in a mixed-beam treatment schedule. At present no therapeutic gain for long-lasting survival has been achieved; however, local control and pain improvement seems to be better with neutrons than with photons.

Introduction

The pelvis, either alone or in combination with other sites, is the earliest and most predominant site of first recurrence of rectal carcinoma. Perineural and venous invasion, tumor size, location, and lymph node status play a very important role as prognostic indicators and are associated with poor locoregional outcome (Athlin et al. 1988; Gerard et al. 1985). Even after improved surgical techniques and adjunctive radiochemotherapy, up to 25% of patients develop local recurrence in the pelvis (Frykholm et al. 1993; Goldmann and Nicholls 1995; Krook et al. 1991). Analyses of some individual surgeons have shown even lower rates (Heald and Ryall 1986).

Recent Results in Cancer Research, Vol. 150

Only a few therapeutic options are available for patients with locally advanced inoperable or recurrent rectal tumors. Traditional palliative local tumor resection has not been very successful (Herfarth et al. 1985; Rominger et al. 1985). The use of combined local hyperthermia and external beam therapy, with or without surgery, has resulted in good palliative responses; however, cure is rare in this patient group. The results of chemotherapy alone have been disappointing, with remission rates of only 10%–20%. More recently, multi-agent chemotherapy has been used concurrently with radiation therapy in such patients (Moertel et al. 1994). Previous studies with photon irradiation for recurrent carcinoma of the rectum reported symptomatic response rates from about 50% to 90%, but the duration of response lasts usually only about 6 months (Flentje et al. 1988; Overgaard et al. 1984; Pacini et al. 1986) and the symptom-free interval is often only one-third of the length of overall survival. Ongoing studies look for the efficacy of intraoperative electron boost therapy following pre- or postoperative external beam radiotherapy (Eble et al. 1995; Gunderson et al. 1983; Lanciano et al. 1993). However, despite all previous efforts of multidisciplinary treatment, the overall prognosis is not favorable and only 15% of all patients with inoperable primary disease and up to 5% with inoperable recurrent disease survive longer than 5 years.

There are various reasons for the therapeutic use of neutrons, which belong to the group of highly ionizing radiation and thus possess a higher linear energy transfer (LET). Advanced inoperable and recurrent rectal cancer have often been observed to be more radioresistant. This increase in radioresistance may be due to hypoxic cell fractions that are identifiable in almost 100% of these tumors or to other intrinsic factors; in any case the use of high-LET radiation therapy may offer a particular advantage (Field and Fowler 1984; Peak et al. 1991). Although initial studies indicate that these tumors may respond better to neutron irradiation, only data from about 350 patients with this indication have been published world wide during the last 20 years (Battermann et al. 1981; Breteau et al. 1986; Caderao et al. 1976; Duncan et al. 1987; Eising et al. 1990; Engenhart et al. 1990, 1992).

Clinical Results

The present review shows results of clinical neutron studies using fast neutrons alone or as boost for the treatment of patients with locally advanced inoperable or recurrent rectal cancer. The review will focus on local control, palliative control, length of survival, and acute and late complications by measurements of clinical and radiological examinations at the follow-up evaluation.

Randomized trials of patients with locally advanced or recurrent rectal cancer have been reported from Edinburgh and Amsterdam (Batterman et al. 1981; Duncan et al. 1987). In these trials neutron beams alone were compared with photons.

In Edinburgh 67 patients were treated, 36 belonging to the advanced inoperable group and 31 to the inoperable recurrent group. For inoperable locally advanced primary rectal cancer, the tumor control rates were similar, with 10% for neutrons and 6.3% for photons. For recurrent tumors the local control rate appeared to be slightly higher, with 33.3% for neutrons and 18.8% for the photon group, but this was not significant because of the small patient number. However, the 3-year survival was around 10% for both treatment modalities. Late severe morbidity was observed mainly in the skin and in hip joints, which was higher in the patient group treated with neutrons (Duncan et al. 1987).

In Amsterdam two studies with different neutron doses were conducted. Results of the pilot study with doses ranging from 17.0 to 25.0 Gy demonstrated steep dose-response curves for tumor control as well as for severe late morbidity (Battermann et al. 1981). In the following study the neutron group was randomized into a low-dose group receiving a total dose of 17.0 Gy and a high-dose group receiving 19.0 Gy. The neutron dose was delivered in 20 fractions over 4 weeks using a six-field technique. This was compared with 50.0 Gy in 20 fractions of photon irradiation.

With 17.0 Gy there was no trend towards improved local/regional control with neutron radiation compared to low-LET photon radiation. However, the tumor control rate increased rapidly from 7% to 38% in the neutron-treated group, which received the higher dose of 19.0 Gy. As a result of severe late neutron effects and some neutron-associated deaths, there was no improvement in survival. All patients subsequently died, 12 from local disease, 2 from intercurrent diseases, but 10 from neutron-induced complications (Battermann 1982).

The overall results available from these randomized studies for the treatment of locally advanced inoperable primary or recurrent rectal cancer with neutrons are discouraging. Both studies showed that the severe late morbidity (proctitis dysuria, hemorrhage, necrosis and perforation) was much more pronounced in the neutron arm than with photons. This high complication rate reflects inherent problems with a suboptimal beam penetration. However, the prospects of superior local tumor clearance rates of 33.3% and 38% reported for neutrons compared to 18% and 10% for photons would be encouraging if the morbidity and mortality could be reduced. Therefore, subsequent studies started to use neutrons only complementary for the small boost volume using muliple-field technique or arc therapy (Breteau et al. 1986; Eising et al. 1990; Engelhart et al. 1989).

In Orléans 58 patients with primarily inoperable or recurrent rectal adenocarcinoma were included in a trial where a neutron boost of 8.3 Gy (N+χ) after an initial series of 40 Gy photons was delivered. This group was compared with a series of 52 patients treated at the Hopital Tenon in Paris. The first part of the treatment was identical: 40 Gy photons was delivered to the entire pelvis in 20 fractions instead of the neutron boost, and 25 Gy photons was given to the reduced volume. Data from 31 neutron-treated patients and 50 photon-treated patients for whom the follow-up period of at least 14 months was available were published. The distribution between inoperable tumors and recurrences was similar in both series (16 inoperable and 15 recurrences in the neutron group vs 24 inoperable and 26 recurrences in the photon group).

Immediate and persistent local tumor control rate for the neutron group at 74% (23/31 patients) and 45% was much higher than that obtained by irradiation with photons alone, 28% (14/50 patients) and 14%. These results reveal a significant difference in local control in favor of the neutron group, with $P<0.01$ at 1 year and $P<0.05$ at 2 years. However, although the survival analysis revealed that there was no difference, one might appear because more patients in the neutron group developed metastases. No differences in immediate and late tolerance were seen.

Between 1976 and 1985, 28 patients with rectal cancer recurrences were treated at the University Hospital of Hamburg. This group consisted of 14 women and 14 men. The average age was 59 years, varying from 53 to 79 years. All patients had primary tumor surgery. Five had had previous irradiation with photons. The indication for neutron irradiation in all cases was unresectable recurrent tumor. Twelve patients complained of uncontrolled severe pain before radiation treatment. Twelve patients received neutron irradiation alone with a median dose of 10.9 Gy, ranging from 7.8 to 14.3 Gy. The mixed beam treated patients received photon doses between 20 and 45 Gy and neutron doses between 5 and 12 Gy with a median dose of 6.9 Gy.

Significant pain relief was reported from 67% of patients and complete pain remission from 17%. This indicates a total pain relief rate of 84%. The median time to progression was 9.5 months, ranging from 0.5 to 47 months. The local control rate at 2 years was 36%. The median survival was 18 months, ranging from 2 to 100 months. In 6/28 patients distant metastases were the reason for death.

Between 1985 and 1995, 44 patients with inoperable recurrences of rectal cancer were treated in Münster. The average age was 66 years, varying from 42 to 72 years. The Karnofsky index averaged 70%, varying from 50% to 90%. In the course of the study 14/44 patients received a second surgical palliative intervention due to the local tumor recurrence. Eight patients had previous photon irradiation with doses between 45 and 60 Gy. Four had residual tumor remaining in the pelvis after attempted primary resection and four were treated preoperatively with photons to render the recurrence tumor resectable. Sixteen of patients also presented with distant metastasis.

Uncontrolled pelvic pain was the reason for neutron radiotherapy in 34 patients (77%). Using CT and MRI visualizing techniques, local infiltration of the urinary bladder was found in eight patients, infiltration of the os sacrum in ten patients, infiltration of the pelvic muscle in six patients, infiltration of the prostate gland in two patients, infiltration of the vagina in three patients and infiltration of the uterus in two patients.

A mixed-beam schedule consisting of 30-45 Gy photons and 5–10 Gy fast neutrons was received by 27/44 patients. The photon therapy was applied by a 10-MV linear accelerator, the neutron therapy by a 14-MeV dt-generator.

Neutron therapy alone was given in 17 patients; the applied dose was between 10 and 15 Gy neutrons (6 patients received a dose of 10 Gy, 1 patient of 13 Gy and 6 patients of 15 Gy). This group also included the four patients who had undergone primarily photon radiotherapy. The neutron radiotherapy consisted of three treatments per week with a single standard dose of 1.67 Gy. The photon radiotherapy was performed on a daily basis. The treatment planning was based on a computer-assisted program using CT and MRI data. An isocentric beam arrangement was used for the neutron and photon radiotherapy, in which the four-field-box technique was used most frequently.

The medium evaluation period was 14 months, with a minimum of 2 months and a maximum of 30 months. There were regular revaluations on a 3 monthly basis.

Significant or total pain relief was achieved in 89% (39/44 patients). A comparison of pain relief between Münster and Heidelberg is shown in Fig. 1. The average relief period was 5.6 months. However, for the high neutron dose group the clinical response shows pain relief at 9 months of 55% compared to 25% for the low neutron dose group. Applying Kaplan-Meier evaluation methods, the statistical probability of no tumor progression at 1 year was 56.4% and the 5-year progression-free survival rate was 9.2%. Median survival was 13.4 months with a 1-year survival rate of 53% and a 5-year survival rate of 3.4%. There was no significant difference between the mixed-beam schedule group and the group treated with neutrons alone.

There was a statistically significant difference with regard to tumor response ($P<0.03$), general health ($P<0.04$) and lymph node affection ($P<0.03$). Those patients who showed partial remission had an average life expectancy of 19 months. Patients with no further progression after treatment had a medium life expectancy of 12.8 months. The average life expectancy in patients with tumor progression was 6.0 months. The 1-year survival rates in patients with partial tumor remission, without tumor progression and with progression were 66.1%, 61.2% and 22.2%, respectively. A 5-year survival probability rate could only be determined for the patients with partial remission. This rate was 3.9% and showed a statistically significance

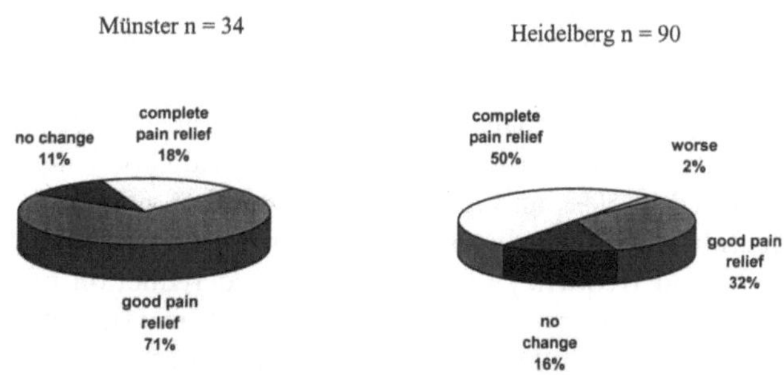

Fig. 1. Comparison of pain relief between Münster and Heidelberg

from the other groups (=<0.04). Patients who were in good health (Karnofsky index, 90%) showed an average survival time of 23.6 months, a 1-year survival rate of 66.7% and a 5-year survival rate of 16.8%. Patients with a Karnofsky index of 80% had an average survival time of 13.8 months and a 1-year survival rate of 57.3%. For patients with a Karnofsky index of 70% or less, the average survival time was 6.3 months and the 1-year survival rate dropped to 22.0%. For the latter two groups it was not possible to determine any 5-year survival rates. This result is statistically significant (P<0.003). Patients with lymph node metastases had an average survival time of 11.3 months and a 1-year survival rate of 33.3% compared to patients without lymph node metastases, who lived 16.8 months on average. This result is also statistically significant (P< 0.02).

A prospective study was conducted in 90 patients with unresectable recurrence of adenocarcinoma irradiated between 1986 and 1993 at the German Cancer Research Center in Heidelberg. All patients were treated with combined photon neutron therapy. The treatment consisted of three-field photon irradiation to the whole pelvis up to a dose of 40 Gy in 20 fractions. In addition, a neutron boost of 6.6 or 10 Gy depending on the presence of systemic disease was applied. The fraction size for the 14 MeV neutrons was 1.1 or 1 Gy given three times a week. For neutrons, arc therapy was planned for all patients. The advantages of this treatment technique have been published previously (Engenhart et al. 1989). No radiation therapy had been given previously. Pain in the pelvic or in the sacroperineal area was the dominant symptom in all patients. The elapsed time between treatment of primary disease and the beginning of radiotherapy treatment was 20 months, with a range from 4 to 48 months. Patients ranged from 41 to 80 years of age (average, 60 years) and the male: female ratio was 1:1. The Karnofsky index was at least 60%. At the beginning of therapy, 30/90 patients had distant metastases, which was the decisive prognostic parameter. Treatment response and duration were anlayzed clinically, with respect to tumor volume on CT or MRI, and laboratory examinations, and in 21 patients the results were correlated with the results of metabolic activity of positron emission tomography (PET) (Engenhart et al. 1992). Response was evaluated clinically in terms of relief of symptoms and decrease in use of analgesics. To obtain an objective measurement of pain, we used the EORTC/RTOG scoring system, which is independent of analgesic therapy (Tong et al. 1982).

Relief of pain was achieved in 98% of patients after radiotherapy. A complete clinical response was defined as a complete relief of pain without the need for further analgesics. Patients with no change still had symptoms but had decreased pain intensity and pain medication. Complete pain relief was achieved in 50%, and a considerable relief of symptoms with a significant reduction of pain medication in 32%. In 16% a only slight improvement in violence and frequency of pain, but a good reduction of pain medication, was achieved. This symptomatic improvement was maintained for 6 months in 78%, for 12 months in 72%, and for 18 months in 51% of the patients (significant at p<0.002). Figure 2 compares pain relief during the 1-year follow-

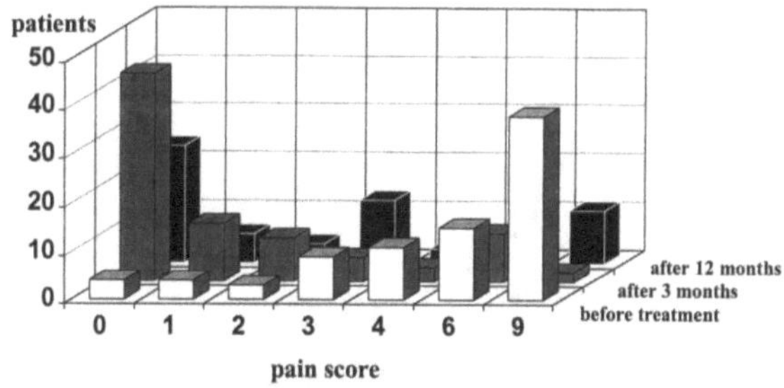

Fig. 2. Pain symptomatic response. Violence of pain at the time of therapy and symptomatic response 3 and 12 months after combined photon-neutron therapy

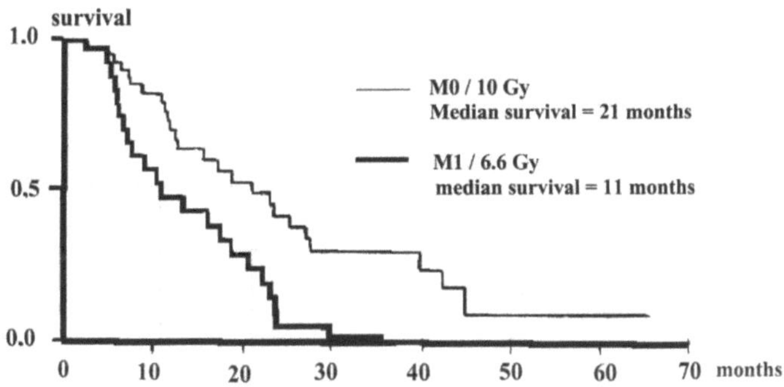

Fig. 3. Overall survival according to systemic metastases

up. The reason for the increase in pain and pain medication 4–31 months after radiotherapy was a late radiation-induced reaction. Eight patients developed proctitis grade III and three proctitis with fistula grade IV.

The median disease-free survival according to Kaplan-Meier methods was 13.9 months, and the median overall survival 15.4 months. For patients with metastases the median survival was 10.6 months and in patients without remote metastases at treatment time it was 21 months (Fig. 3). The 2 and 5 year survival rates were 3% and 0% compared to 42% and 8% for patients without metastases. A further factor that affected the prognosis of patients with inoperable recurrent carcinoma was the health status at time of treatment. The median survival for patients with a Karnofsky performance index of ≥ 90% was 21.5 months compared to 14 months for those with an index of 70%–80% and only 6.5 months for the group of patients with a Karnofsky index of 50%. The influence of Karnofsky performance index is shown in Fig. 4. Table 1 summarizes the pain relief and local control rates according to the Kaplan-Meier method at four different European centers. These patients were irradiated with a mixed - beam schedule consisting of 40 Gy photons (30–45 Gy) delivered to the whole pelvis and a neutron dose of 5–10 Gy as boost. In the Münster group, 17 patients and in the Hamburg group 12 patients were treated with neutrons alone.

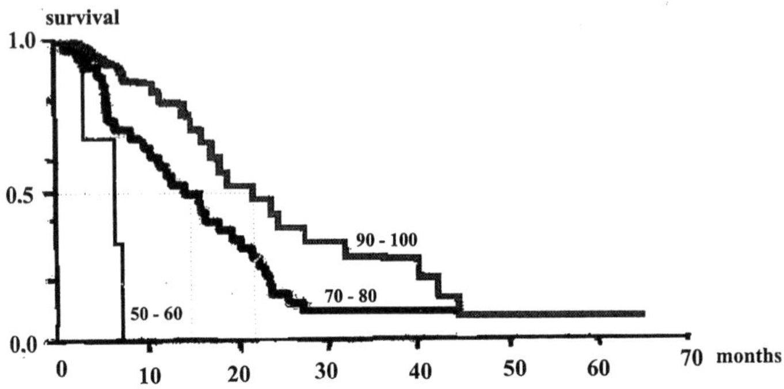

Fig. 4. Survival rate of patients with recurrences of rectal cancer according to Karnofsky status

Table 1. Local control rates at four centers according to life-table method and pain relief for inoperable or recurrent rectal cancer

Institution	Indications	Patients	Photons/neutrons (%)		Pain relief
			LCR (1a)	LCR (2a)	
Orléans	Nonresectable primary and recurrences	31	60	38	–
Münster	Recurrences	37	56	26	89%
Heidelberg	Recurrences	90	56	40	98%
Hamburg	Nonresectable primary and recurrences	28	–	36	84%

Conclusions

The overall survival and quality of life following pelvic recurrence is dismal and the clinical course of patients with pelvic recurrence is characterized by extreme morbidity. In view of these findings and based on biological knowledge, it has been suggested that patients with inoperable recurrent rectal cancer may benefit from high-LET irradiation.

The randomized trials of advanced inoperable or recurrent adenocarcinoma of the rectum showed an advantage of neutrons over photons with respect to control rates but not with respect to survival. Side effects were statistically higher and patients died as a result of late radiation complications (Battermann 1982; Duncan et al. 1987). The high complication rate reflects inherent problems with the poor penetration obtained with fixed DT generators or low-energy cyclotrones, which are much worse than those achieved with photons and the known narrow therapeutic ratio for late sequelae (Cohen et al. 1995; Errington et al. 1991).

In the study from Heidelberg, Münster and Orléans, the survival curves were not significantly different (Fig. 5). At 1 year the survival rates were between 53% and 70%; after 2 years the survival rates ranged between 32% and 40%. In the Orléans group a plateau of survival seemed to exist at

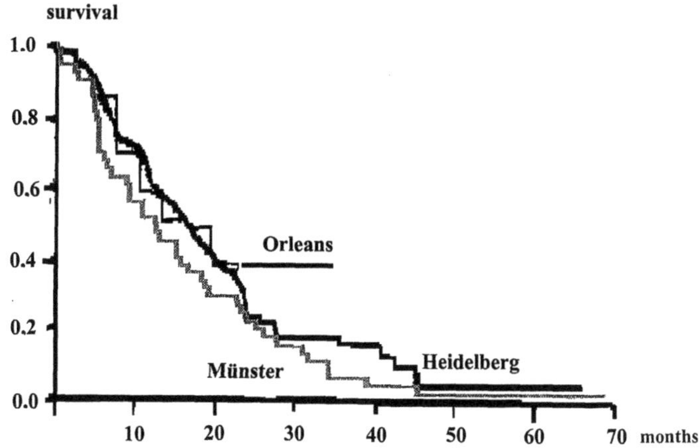

Fig. 5. Survival comparison of three different centers

2 years. In this study, 16/31 patients were suffering from inoperable rectal cancer and only a minority had distant metastases. In the Heidelberg and Münster groups no patient with inoperable rectal cancer but a high proportion of patients with distant metastases were included; these are known to have a poor prognosis. In the Heidelberg study, the 2-year survival rate for patients with distant metastases was only 3% compared to 42% for patients without metastases; at 5 years the survival rates were 0% and 8%.

In spite of the general unfavorable condition of patients with large tumor extension, the palliative effect in terms of pain relief was high, in the range of over 80%. In the Heidelberg group, the symptomatic improvement was maintained for 6 months in 78%, for 12 months in 69% and for 18 months in 51% of the patients. This is advantageous compared to the results reported in the literature for patients exclusively treated by photon therapy. Following photon therapy, symptomatic response rates range from 50% to 90%, with a duration of 12 months in 10%–30%. Local failure is the usual cause leading to a limited duration of pain relief. The probability of relapse-free survival was 56% in Münster and 56% in Heidelberg after 1 year, with a median disease-free survival of 13.4 months for Münster and of 15.4 months for Heidelberg (Prott et al. 1997).

There were only few acute side effects that were comparable to those in the photon-treated group. In the Heidelberg group, 16 patients developed moderate or severe late complications. We saw eight grade III and three grade IV small bowel complications. Eleven of 16 patients received 10 Gy neutrons in addition to 40 Gy photons. The higher complication rate can be caused by the higher dose or the longer survival time. However, this late severe normal tissue toxicity appeared to be similar to that of a historical photon group of 74 patients treated in Heidelberg with a dose of 60–66 Gy (Flentje et al. 1988). In the Orléans group, there was also no significant difference in late complications. Only pelvic sclerosis was more frequent in the mixed photon-neutron group than in the historical photon group (9/31 vs 11/50) (Breteau et al. 1986).

To date there has been no clear clinical indication that fast neutron irradiation alone or in a mixed schedule can produce better results than photon irradiation in the treatment of patients with rectal adenocarcinoma. However, the number of patients treated is small and, in addition, the penetration of the neutron beams was suboptimal for the two randomized trials. By using sophisticated treatment techniques and neutron beams produced by protons of 45 MeV or more, optimal homogeneous dose distribution in the tumor volume with a sufficient dose reduction in cutaneous and subcutaneous tissue, bladder and small bowel can be achieved. Taking into account the strong pain relief which improves quality of life and the high local control rates, neutron therapy affords an opportunity for treatment, especially for non metastatic locally extended, inoperable or recurrent tumors.

These results compare very favorably with the results published of intraoperative irradiation after palliative surgery for locally recurrent rectal cancer. The Mayo Clinic experience in 106 patients who underwent palliative resections indicated 3- and 5-year survivals of 28% and 12%. Outcome depends on surgical status. Patients with only microscopic residual disease appeared to have a better survival: 33% compared with 9% with gross residual disease. Complications were observed in 36% of patients related to pelvic cancer and small bowel. The complication rates with neutron treatment may be tolerable in relation to the higher local control rates. For adjuvant radiotherapy, however, these complication rates might not be tolerable. However, we have to keep in mind that these results are nowadays in competition with those of other treatment modalities such as radiochemotherapy, hyperthermia, intraoperative radiotherapy and brachytherapy (Eble et al. 1995; Gunderson et al. 1983; Lanciano et al. 1993).

References

Athlin L, Bengtsson NO, Stenling R (1988) Local recurrence and survival after radical resection of rectal carcinoma. Acta Chir Scand 154:225–229

Battermann JJ, Hart GAM, Breur K (1981) Dose-effect relations for tumour control and complication rate after fast neutron therapy for pelvic tumors. Br J Radiol 54:899–904

Battermann JJ (1982) Results of dT fast neutron irradiation on advanced tumors of bladder and rectum. Int J Radiat Oncol Biol Phys 8:2159–2164

Breteau N, Destembert B, Favre A, Sabattier R, Schlienger M (1986) An interim assessment of the experience of fast neutron boost in inoperable rectal carcinoma in Orléans. Bull Cancer/Radiother 73:591–595

Caderao JB, Hussey DH, Fletcher DH, Sampiere VA, Johnson DE, Wahrton JT (1976) Fast neutron radiotherapy for locally advanced pelvic cancer. Cancer 37:2620–2629

Cohen L, Saroja KR, Hendrickson FR, Lennox AJ, Hatcher MA, Kroc TK (1995) Neutron irradiation of human pelvic tissues yields a steep dose-response function for late sequelae. Int J Rad Oncol Biol Phys 32:367–372

Duncan W, Arnott SJ, Jack WJL, Orr JA, Kerr GR, Williams JR (1987) Results of two randomised clinical trials of neutron therapy in rectal adenocarcinoma. Radioth Oncol 8:191–198

Eble M, Wulf J, van Kampen M, Weischädel U, Wannenmacher M (1995) Lokal begrenzte Dosiseskalation in der Radiotherapie des primär ausgedehnten und lokal rezidivierten Rektumkarzinoms. Strahlenther Onkol 171:77–86
Eising E, Pötter R, Haverkamp U, Schnepper E (1990) Neutron therapy (dT, 14 MeV) for recurrents of rectal cancer: interim analyses from Münster. Strahlenther Onkol 166:90–94
Engenhart R, Kimmig B, Marin Grez M, Höver KH, Flentje M, zum Winkel K (1989) Combined neutron photon therapy of locally recurrent rectosigmoidal tumors. Strahlenther Onkol 165:327–329
Engenhart R, Kimmig B, Wannenmacher M, Höver KH, Hesse BM, Lorenz JW (1989) Experience with the computer controlled weighted moving beam therapy (WMBT) in routine clinical use. In: Blattmann H (ed) Proceedings of the International Heavy Particle Therapy Workshop (PTCOG/EORTC/ECNEU), Villingen 1989, pp 38–42
Engenhart R, Kimmig B, Höver KH, Strauss LG, Lorenz WJ, Wannenmacher M (1990) Photon-neutron therapy for recurrent colorectal cancer-follow up and preliminary results. Strahlenther Onkol 166:95–98
Engenhart R, Kimmig B, Strauß G, Höver KH, Rohman J, Haberkorn U, van Kaick G, Wannenmacher M (1992) Therapy monitoring of presacral recurrences after high-dose irradiation: value of PET, CT, CEA and pain score. Strahlenther Onkol 4:203–212
Errington RB, Ashby D, Gore SM, Abrams KR, Myint S, Bonnett DE, Blake SW, Saxton TE (1991) High energy neutron treatment for pelvic cancers: study stopped because of increased mortality. BMJ 302:1045–1051
Field SB, Fowler JF (1984) The biological basis for neutron therapy. J Eur Radiother 3:170–175
Flentje M, Frey M, Kuttig H, Kimmig M (1988) Strahlentherapie bei Lokalrezidiven colorektaler Tumoren. Prognostische Faktoren, Verlaufsdiagnostik und Ergebnisse. Strahlenther Onkol 164:402–407
Frykholm G, Glimelius B, Pahiman L (1993) Pre- and postoperative irradiation in adenocarcinoma of the rectum. Final treatment results of a randomized trial and an evaluation of late secondary effects. Dis Colon Rectum 36:564–572
Gerard A, Berrod JL, Pene F, Loygue J, Langier A, Bruckner R, Camelot G, Arnand JP, Metzger U, Buyse M, Dalesio O, Duez N (1985) Interim analysis of a phase III study on preoperative radiation therapy in resectable rectal carcinoma. Cancer 55:2373–2379
Goldmann PA Nicholls RJ (1995) Prediction of local recurrence and survival of carcinoma of the rectum by surgical and histopathological assessment of local clearance. Br J Surg 82:1054–1056
Gunderson L, Cohen AC, Dosoretz DD, Ship WU, Hedberg SE, Wood WC, Rodkey GV (1983) Residual, unresectable, or recurrent colorectal cancer: external beam irradiation and intraoperative electron beam boost - resection. Int J Radiat Oncol Biol Phys 9:1597–1603
Heald RJ, Ryall RDH (1986) Recurrence and survival after total mesorectal excision for rectal cancer. Lancet 2:1479–1482
Herfarth Ch, Schlag P, Hohenberger P (1985) Möglichkeiten bei locoregionären Rezidiven der Carcinome des Gastrointestinaltraktes. Chirurg 56:492–498
Krook JE, Moertel CG, Gunderson LL et al (1991) Effective surgical adjuvant therapy for high-risk rectal carcinoma. N Engl J Med 324:709–715
Lanciano RM, Calkins AR, Wolkov HB, Budzydlowski J, Noyes RD et al (1993) A phase I/II study of intraoperative radiotherapy in advanced unresectable or recurrent carcinoma of the rectum: a RTOG study. J Surg Oncol 1:20–29
Moertel CG, Gunderson LL, Mailliard JA, McKenna PJ, Martenso JA, Burch PA, Cha SS (1994) Early evaluation of combined fluorouracil and leucovorin as a radiation enhancer for locally unresectable, residual, or recurrent gastrointestinal carcinoma. The North Central Cancer Treatment Group. J Clin Oncol 1:21–27
Overgaard M, Overgaard J, Sell A (1984) Dose response relationship for radiation therapy of recurrent, residual and primarily inoperable colorectal cancer. Radiother Oncol 1:217–225

Pacini P, Cionini L, Pirtoli L, Ciatto S, Tucci E, Sebaste L (1986) Symptomatic recurrences of carcinomas of the rectum and sigmoid. The influence of radiotherapy on the quality of life. Dis Colon Rect 29:865–868

Peak MJ, Wang L, Hill CK, Peak JG (1991) Comparison of repair of DNA double-strand breaks caused by neutron or gamma radiation in cultured human cells. Int J Radiat Biol 60:891–898

Prott FJ, Pötter R, Preusser PO, Micke O, Schouwing A, Willich N (1997) Ergebnisse der Behandlung von Rektumkarzinomrezidiven mit schnellen Neutronen (dT 14 MeV). Strahlenther Onkol 173:316–322

Rominger CJ, Helber RD, Gunderson LL et al (1985) Radiation therapy alone or in combination with chemotherapy in the treatment of residual or inoperable carcinoma of the rectum and rectosigmoid or pelvic recurrence following colorectal surgery: Radiation Therapy Oncology Group study (76-16). Am J Clin Oncol 8:118–127

Tong D, Gillick L, Hendrickson FR (1982) The palliation of symptomatic osseous metastases. Cancer 50:893–899

Fast Neutrons in Prostatic Adenocarcinomas: Worldwide Clinical Experience

K. L. Lindsley, P. Cho, K. J. Stelzer, W.-J. Koh, M. Austin-Seymour, K. J. Russell, G. E. Laramore, and T. W. Griffin

Department of Radiation Oncology, University of Washington Medical Center, 1959 NE Pacific Street, Seattle, WA 98195, USA

Abstract

Primary tumor control remains a major problem in the treatment of locally advanced prostate carcinoma. Clinical local failure rates approach 30–40% and may be significantly higher when results of prostatic biopsy or prostate-specific antigen (PSA) levels are considered. The low growth rate and cycling fraction of prostate adenocarcinoma suggest potential therapeutic advantage for the high linear energy transfer (LET) of neutrons. The Radiation Therapy Oncology Group (RTOG) performed a multi-institutional randomized trial (RTOG 77-04) comparing mixed beam (neutron plus photon) irradiation to conventional photon irradiation for the treatment of locally advanced prostate cancer. A subsequent trial by the Neutron Therapy Collaborative Working Group (NTCWG 85-23) compared pure neutron irradiation to standard photon irradiation. Both randomized trials demonstrate significant improvement in locoregional control with neutron irradiation compared to conventional photon irradiation in the treatment of locally advanced prostate carcinoma. To date, only the mixed beam trial has shown a significant survival benefit. Future analysis of the larger NTCWG trial at the 10-year point should confirm whether or not improved locoregional control translates into a survival advantage. These findings have significant implications for all local treatment strategies including dose-escalated conformal photon irradiation, prostate implantation, and neutron radiation. Given the large numbers of patients afflicted with this disease, a positive survival advantage for neutrons or mixed beam therapy would provide a strong incentive for the development of economically feasible clinical neutron facilities.

Introduction

Excluding skin cancer, prostate carcinoma is the most common cancer in men in the United States and Europe, and is second only to lung cancer as the leading cause of cancer death (Wingo et al. 1995). Radiotherapy has been the mainstay of treatment for locally advanced T3 and T4 prostate tumors;

Recent Results in Cancer Research, Vol. 150

however, over the last few decades, the shortcomings of radiotherapy in achieving reliable tumor sterilization have been increasingly recognized. Local failure assessed clinically occurs in 30%–40% of locally advanced patients (Duncan et al. 1993, Hanks et al. 1987), and may be significantly higher if rising prostate specific antigen (PSA) or results of prostatic biopsy are considered (Freiha and Bagshaw 1984; Stamey et al. 1993).

The high linear energy transfer (LET) of neutrons offers a potential radiobiological advantage over photon irradiation in overcoming tumor radioresistance in slowly cycling and/or advanced hypoxic tumors. The low growth rate and cycling fraction of prostate adenocarcinoma made this tumor a logical target for trials comparing neutron beam radiation to conventional photon treatment. Initial clinical trials of neutron radiation for the treatment of pelvic tumors were hampered by the poor depth dose penetration and limited beam-shaping capabilities of early laboratory-based facilities. The development of clinical neutron facilities in the 1980s, capable of delivering higher energy neutrons, spurred full-scale phase III testing of neutron beam radiotherapy for several distinct tumors including adenocarcinoma of the prostate.

Randomized Trials of Neutron Radiation

The first randomized trial of neutron radiation for the treatment of prostate cancer was initiated by the Radiation Therapy Oncology Group (RTOG). The study (RTOG 77-04) employed a mixed beam (2×/week neutrons, 3×/week photons) for the experimental treatment (Laramore et al. 1985, 1993). The neutron/photon combination was selected over a pure neutron beam approach due to the inferior depth dose penetration of neutrons and concerns for enhanced late effects with neutrons. A total of 91 evaluable patients with biopsy proven stage C or D_1 adenocarcinoma of the prostate were entered between June 1977 and April 1983, and randomized between mixed-beam versus conventional photon irradiation. The randomization was purposefully unbalanced (60%–40%) with more patients randomized to the experimental mixed-beam arm (55 patients) than to the photon group (36 patients). Standard photon treatment consisted of whole pelvis radiation to a dose of 50 Gy at a fractionation of 1.8–2 Gy/day, followed by a 20-Gy boost to the prostate and any areas of gross extraprostatic disease. Mixed-beam radiation was delivered to the same photon-equivalent dose and fractionation, with an adjustment in the neutron daily dose based on the relative biological effectiveness (RBE) of the treating institution which ranged from 3.0 to 3.3. The experimental group received neutron treatments 2 days/week and photons 3 days/week.

The 10-year results of RTOG 77-04 were reported by Laramore et al. (1993). Figure 1 shows the proportion of patients maintaining clinical local-regional control as a function of time. Clinical local-regional failure was defined as

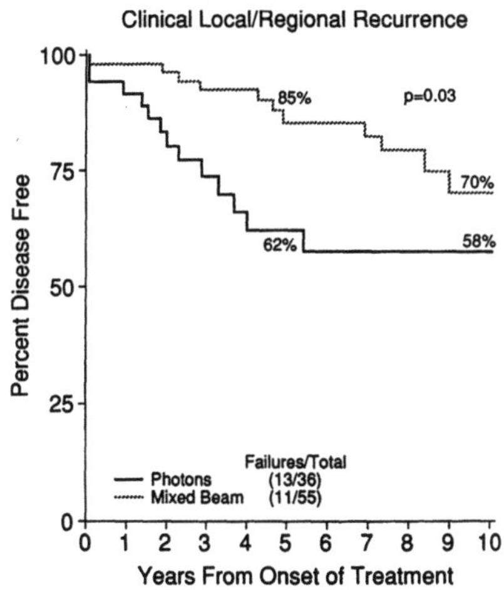

Fig. 1. Clinical local-regional control in patients with locally advanced prostate cancer treated on RTOG 77-04. (Laramore et al. 1993)

- a 25% or greater increase in the product of the tumor major diameters assessed by digital rectal exam,
- new extension of tumor locally or beyond the prostate capsule or re-extension after becoming temporarily negative,
- new development of positive nodes or reappearance of positive nodes after becoming temporarily negative, or
- clinical evidence of local tumor progression, such as obstructive symptoms, confirmed by positive biopsy.

At 10 years, the clinical local-regional control rate was 70% for the mixed-beam group compared to 58% for the standard photon arm. The difference between the curves is statistically significant ($p=0.03$). As part of a separate program to assess local tumor control histologically, 11 patients from the University of Washington who were clinically free of local-regional progression with normal serum acid phosphatase underwent random biopsies (2 from each lobe) of the prostate. Patient sampling bias occurred since more than twice as many mixed-beam-treated patients were alive at the time the biopsies were being done. Positive biopsies were obtained in four mixed-beam patients and one photon patient. All five have remained clinically free of local progression for over 5 years. Inclusion of these "pathologic only" failures in the analysis yielded local control rates at 10 years of 61% for mixed-beam patients and 52% for photon patients.

Actuarial survival for the two groups of patient are shown in Fig. 2. A significant ($p=0.04$) survival advantage was seen for neutrons at both the 5-year (70% vs. 53%) and 10-year (46% vs. 29%) end points. Skin and subcutaneous tissue reactions were more severe in the mixed-beam group, due to the inferior penetrating properties of neutron beams. However, other sequelae including gastrointestinal, rectal, and urinary were similar. There was no significant difference in the incidence of severe toxicity, which occurred in

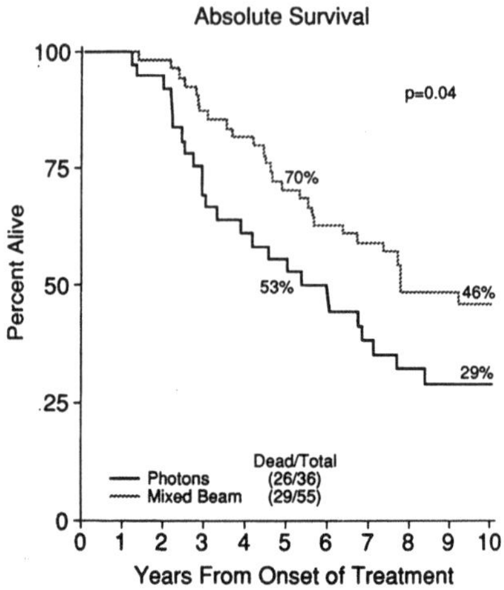

Fig. 2. Absolute survival of patients with locally advanced prostate cancer treated on RTOG 77-04. (Laramore et al. 1993)

seven mixed-beam-treated patients and five photon-treated patients, including one fatal complication in each arm.

Although the study demonstrated statistically significant improvement in both local-regional control and survival for the mixed-beam therapy, considerable skepticism remained regarding the superiority of neutrons over photons in the treatment of advanced prostate carcinoma. The total number of patients enrolled in the trial was relatively small. Critics pointed to the relatively poor local control and survival of patients in the photon arm compared to other previously reported series of similarly staged patients treated with photons. However, most other series did not include stage D_1 patients (which accounted for 12% of patients in the RTOG study), and direct comparisons are further complicated by patient selection factors across different institutions.

Encouraged by the preliminary results of the RTOG study, the Neutron Therapy Collaborative Working Group (NTCWG) began a phase III randomized trial in 1986 comparing fast neutrons to conventional photon radiation for locally advanced prostate carcinoma. Through the National Cancer Institute (NCI)-sponsored development of modern hospital-based cyclotrons, higher energy beams had become available, allowing the feasibility of "pure" neutron-beam radiation for prostate adenocarcinoma. The study (NTCWG 85-23) was closed to patient entry in October of 1990, following the accrual of 172 evaluable patients, including 87 patients randomized to neutrons and 85 patients randomized to photons (Russell et al. 1993). Patients eligible for study were those with stage T3-4N0-1M0 tumors of any histological grade as well as patients with high-grade T2 tumors (summed major and minor Gleason pattern scores >7). Exclusion criteria included prior definitive prostate surgery, androgen ablation therapy, nodal involvement at or above the common iliac nodes, or performance status < 50. Pretreatment staging evaluation included history and physical, complete blood count (CBC), serum acid phosphatase, chest X-ray, plevic CT, and bone scan. Staging pelvic lymphadenectomy was recommended but not required, and was performed in 44% of neutron patients and 42% of photon patients. As PSA assays became routinely available they were obtained pretreatment. Standard photon treatment included pelvic radiation to a total dose of 50–50.4 Gy with 6 MV or higher photons at a daily fractionation of 1.8–2 Gy. The primary prostate tumor was subsequently boosted for an additional 20 Gy for a total dose of 70 Gy. For neutron treatment, the pelvis received a total dose of 13.6 nGy at a fractionation 1.7 nGy delivered 3 times/week. The primary prostate tumor was boosted for an additional 6.8 nGy (same fractionation) for a total dose of 20.4 nGy. Standard four-field box techniques were used for both photon and neutron radiation. Customized shielding of femoral neck, rectum, and bladder was used during photon radiation, but varied greatly between institutions for neutron treatments depending on their beam-shaping capabilities. Primary tumor response was measured clinically by palpation at the time of digital rectal examination. Complete clinical response was defined as complete regression of palpable abnormalities in the prostate or periprostatic tissues. Progression of the primary was defined as a 25% or greater increase in the product of the length and width of the tumor nodule or a positive biopsy in the setting of clinical tumor growth. All local failures were confirmed by positive biopsy. Positive biopsies obtained 2 or more years after treatment for protocol purposes in the absence of clinical evidence of local progression were not scored as a clinical local tumor failure. However, a separate analysis of "histological local-regional failures" was performed which included these "biopsy only failures" in addition to the clinical local-regional failures. Entry level PSA values were obtained consistently only towards the end of the study due to availability; however, post-treatment PSA values were obtained on all patients at risk for 5 years.

High rates of clinical complete response (94% for neutrons and 96% for photons, $p=$n.s.) were obtained in both arms by 2 years after treatment. The clinical local-regional failure rates are shown in Fig. 3. The difference in the curves is statistically significant ($p<0.01$) with projected 5-year clinical local-regional failure rates of 11% for neutrons and 32% for photons. If "biopsy only failures" are included, the 5-year "histological local-regional tumor failure rates remain significantly different (13% vs. 32%). Similarly, "biochemical" control as defined by PSA levels was higher following neutron irradiation. PSA levels at 5 years post-treatment were elevated in 17% of neutron patients compared to 45% of photon-treated patients ($p<0.001$). Despite differences in locoregional control, the overall survival rates by treatment projected to 5 years are equivalent as shown in Fig. 4. At the time of the last published analysis, only 17 prostate-cancer-related deaths had occurred (8 and 9 in the neutron and photon groups, respectively).

Acute toxicity for both treatment regimes was acceptable with no grade 4 or higher toxicity. However, late complications were significantly higher following neutron radiation as shown in Fig. 5. The 5-year actuarial rate of grade 4 late complications was 11% for neutrons compared to 3% for photons. No treatment-related deaths were reported. Large bowel toxicity, ne-

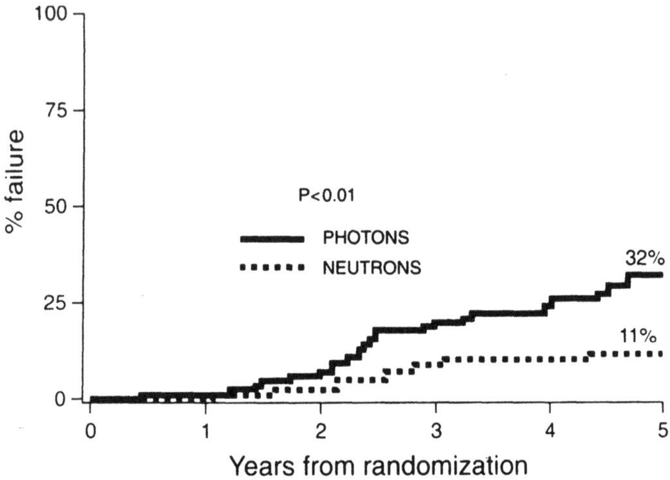

Fig. 3. Actuarial clinical local-regional failure in patients with locally advanced prostate cancer treated on NTCWG 85-23. (Russell et al. 1993)

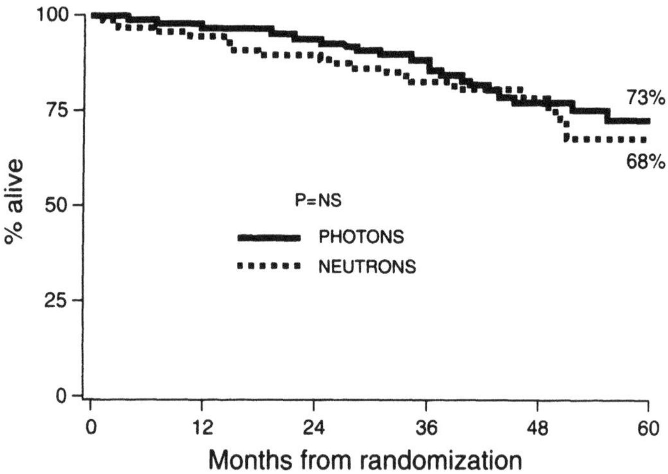

Fig. 4. Actuarial survival of patients with locally advanced prostate cancer treated on NTCWG 85-23. (Russell et al. 1993)

cessitating surgical intervention and colostomies, accounted for the majority of the increase in late sequelae. The incidence of severe late sequelae in the neutron treatment group appeared to be related to the beam-shaping capabilities of the neutron facility. Differences between neutron facilities in the incidence of required colostomies, as well as overall grade 3 and 4 late toxicity, were statistically significant. A subsequent analysis of the data including a separate analysis of the University of Washington experience (Austin-Sey-

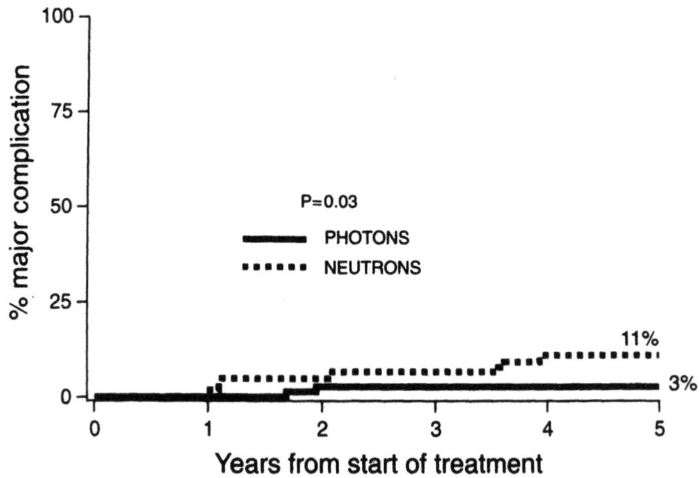

Fig. 5. Actuarial incidence of complications (grade 4) in patients with locally advanced prostate cancer treated on NTCWG 85-23. (Russell et al. 1993)

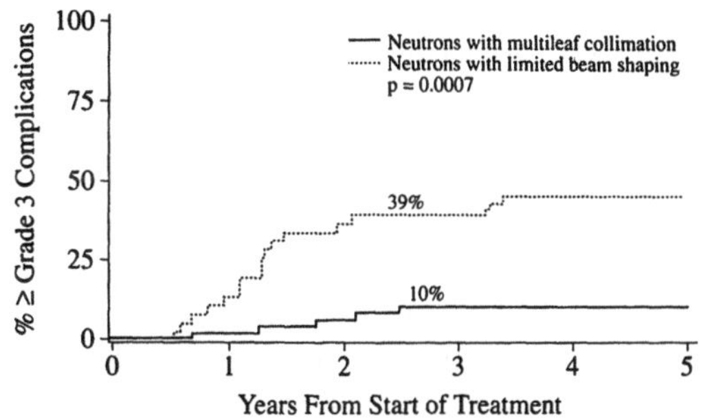

Fig. 6. Cumulative incidence of severe (≥grade 3) late complications for patients treated with neutrons on NTCWG 85-23. The two groups are patients treated with neutrons using multileaf collimation at University of Washington and patients treated with neutrons with limited blocking capabilities at the two other participating institutions. (Austin-Seymour et al. 1994)

mour et al. 1994) confirmed the importance of beam collimation in the incidence of late sequelae. Figure 6 shows the cumulative incidence of severe (grade 3 or higher) late complications for patients treated with neutrons in the randomized trial. Patients treated at the two institutions with limited beam-shaping (M.D. Anderson and the University of California, Los Angeles) experienced a significantly higher complication rate (39%) than those treated at the University of Washington (10%), where continuously variable multileaf collimation was available. Furthermore, an analysis of the 97 patients entered on the randomized trial at the University of Washington showed no significant difference between photon-treated and neutron-treated patients in the incidence of severe late complications (Fig. 7).

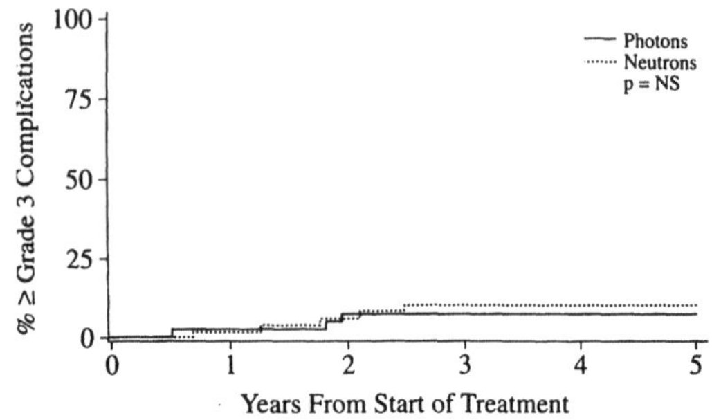

Fig. 7. Cumulative incidence of major (grade 3 or higher) complications in patients treated at the University of Washington with neutrons or photons on NTCWG 85-23. (Austin-Seymour et al. 1994)

Sequelae of Neutron Radiation for Prostate Carcinoma

Skeptics of neutron radiation have argued that the improved local control rates have come at the expense of unacceptable late tissue morbidity (Duncan 1994). Clearly, normal tissue dose response curves for neutron irradiation are steep. Well-collimated, high-enery beams are essential to achieving adequate dose distributions for prostate radiation. A review of the worldwide neutron treatment registry revealed a high severe late normal tissue complication rate associated with low-energy neutrons, not observed with high-energy neutron radiation (Griffin and Laramore 1995; Griffin et al. 1986).

As reflected in the analysis of late tissue complications in the randomized NTCWG trial, adequate beam collimation was critical to the maintenance of an acceptably low toxicity profile for fast neutrons. The importance of treatment planning as well as beam energy were recently confirmed in a report from the University of Chicago, involving 45 patients treated with mixed-beam (40% neutron) radiation for bulky B_2, C, and D_1 stage prostate cancer (Haraf 1995). Neutron radiation was delivered at either the University of Chicago or the Fermi National Accelerator Laboratory, depending on which facility was nearest to the patient's home. While the overall survival and local control were felt to be superior to those series using photon radiation, the rate of serious grade 3–5 complications was unacceptably high at 36%. The site of neutron irradiation was a major factor in the development of late tissue sequelae, with all rectal complications occurring in patients treated at the lower energy Fermilab facility. The Fermilab beam has a depth of maximum dose (d_{max}) of 0.15 cm and depth dose characteristics comparable to a cobalt beam with 70-cm source to skin distance (SSD). The only factor that correlated to the development of long-term bladder complication was field arrangement, with increased toxicity occuring for the four-field oblique technique, which yielded potential hot spots in the bladder.

In contrast, no grade 3–5 gastrointestinal (GI) or genitourinary (GU) complications were reported in 151 patients treated at Wayne State University in three prospective dose-finding studies involving conformal mixed neutron or photon radiation (Chuba et al. 1996; Forman et al. 1995, 1996a). However, hip stiffness was found to be a dose-limiting toxicity in the mixed-beam studies with incidence related to the proportion of dose delivered with neutrons and the volume of normal tissue (gluteus maximus muscle) irradiated.

Increased transient and chronic neurological complications have also been associated with neutron beam radiation for prostate carcinoma (Russell et al. 1990). A review of 132 patients treated on prospective trials at the University of Washington revealed neurological complications in 26/194 patients treated with fast neutrons only, 3/16 patients treated with mixed beam, and 2/22 patients treated with photons only ($p=0.16$). Two distinct neurological syndromes were described. A sciatica syndrome consisting of pain, weakness, and/or numbness of the lower extremities occurred during or 2–7 months after completion of treatment in 24 patients. Most cases were reversible with rapid response to steroid therapy; however, one patient had severe residual lower extremity weakness, numbness, and pain. Electromyography (EMG) and nerve conduction velocity (NCV) studies revealed bilateral peripheral nerve denervation without signs of demyelinization, with the pathologic lesion estimated as distal to the spinal roots. A sphincter dysfunction syndrome, consisting of signs and symptoms of sacral plexus injury, occurred in ten patients, all treated with neutrons alone, 1–15 months post-therapy. Three patients had asymptomatic diminshed rectal tone, six patients had occasional rectal or urinary incontinence (not requiring protective undergarments), and one patient had severe rectal and bladder incontinence with EMG testing revealing an S2–S4 peripheral nerve injury. Pretreatment risk factors included a history of hypertension, diabetes, smoking, or peripheral vascular disease. The neurological complications possibly reflect an higher RBE of fast neurons for peripheral nerves as compared to that of other pelvic normal tissues. Nonetheless, the overall small percentage of severely affected patients treated with neutrons (2/110) was not felt to be prohibitive.

Future Directions

Both randomized trials (RTOG 77-04 and NTCWG 85-23) demonstrate significant improvement in locoregional control with neutron irradiation compared to conventional photon irradiation in the treatment of locally advanced prostate carcinoma. To date, only the mixed-beam trial has shown a significant survival benefit. Improvements in local-regional control alone (without survival benefit) may not be sufficient justification for wide-scale application of neutron beam irradiation, since quality of life is less impacted by local control in this disease. Future analysis of the larger NTCWG trial at the 10-year point should confirm whether or not improved locoregional control translates into a survival advantage. These findings have significant implications for all local treatment strategies including dose-escalated conformal photon irradiation, prostate implantation, and neutron radiation. The association of clinical local failure with subsequent development of distant metastases and decreased survival has been observed in a number of retrospective studies of patients treated with iodine-125 implantation (Fuks et al. 1991) or conventional photon irradiation (Kuban et al. 1989; Zagers et al. 1991). Additionally microscopic residual as assessed by prostatic biopsy following radiation therapy is associated with a significantly increased risk of subsequent clinical local and distant failure (Scardino et al. 1986; Schellhammer et al. 1987). These results suggest that tumor regrowth at the primary site may lead to an enhanced risk of metastatic disease. Of note is the relatively long time interval observed between local failure and clinical expression of metastatic disease, with reported delays of 6.5 years or more in one-half of patients (Fuks et al. 1991).

It remains to be seen whether or not neutron irradiation can achieve superior local control beyond that which can be obtained with conformal dose escalation of photon radiation. The Patterns of Care Studies of conventional photon radiation demonstrated a dose response relationship for patients with stage C cancers with improved local control seen at doses above 70 Gy (Hanks et al. 1985, 1988). Not surprisingly, higher doses also resulted in an increased rate of complications. Improvements in conformal technology and three-dimensional reconstruction of the prostate have allowed further dose escalation of photon irradiation with acceptable toxicity. A report from the Fox Chase Cancer Center showed improvement in biochemial control with dose escalation above 71 and 73 Gy, for patients with pretreatment PSA above 10 ng/ml (Hanks et al. 1996). Biochemical failure was defined as PSA greater or equal to 1.5 ng/ml and rising on two consecutive occasions. No improvement in biochemical control was seen with dose escalation for patients with pretreatment PSA less than 10 ng/ml.

Two series of prospective dose-finding studies have been conducted at Wayne State University involving hyperfractionated conformal photon radiation and conformal neutron radiation for locally advanced prostate cancer. An analysis of results shows similar rates of PSA decline and 12-month negative biopsies for patients treated with a mixed neutron-photon dose of 15 nGy plus 18 PhGy (photon Gray) compared to those receiving photon hyperfractionation to 83 Gy in conjunction with neoadjuvant hormonal therapy (Forman et al. 1996b). No significant difference was seen in the rate of chronic GI or GU toxicity; however, the neutron group experienced a higher rate of moderate (14%) or severe (12%) hip stiffness.

Ultimately, a randomized trial between dose-escalated conformal photon irradiation and conformal neutron irradiation may be required to definitively assess the relative merits of each with regard to locoregional control, survival, and treatment complications. However, much remains to be ascertained regarding the optimization of both modalities with regard to final dose, dose fractionation, treatment volumes, and the use of neoadjuvant or adjuvant hormonal therapy. Furthermore it is possible that optimization of therapy may involve a defined mix of neutron and photon irradiaton. Future trials should limit neutron radiation of facilities with high-energy beams with sophisticated collimation and full rotational capabilities. In addition stratifcation of patients by known prognostic factors including stage and pretreatment PSA (Fukunaga-Johnson et al. 1997; Hanks et al. 1996) is essential. Due to the large numbers of patients afflicted with this disease, a positive survival result for neutrons or mixed-beam therapy would provide a strong incentive for the development of economically feasible clinical neutron facilities.

References

Austin-Seymour M, Caplan R, Russell K, Laramore G, Jacky J, Wootton P, Hummel S, Lindsley K, Griffin T (1994) Impact of a multileaf collimator on treatment morbidity in localized carcinoma of the prostate. Int J Radiat Oncol Biol Phys 30:1065–1071

Chuba PJ, Sharma R, Yudelev M, Duclos M, Shamsa F, Giacalone S, Orton CG, Maughan RL, Forman JD (1996) Hip stiffness following mixed conformal neutron and photon radiotherapy: a dose-volume relationship. Int J Radiat Oncol Biol Phys 35:693–699

Duncan W (1994) An evaluation of the results of neutron therapy trials. Acta Oncol 33:299–306

Duncan W, Warde P, Catton CN, Munro AJ, Lakier R, Gadella T, Gospodarowicz MK (1993) Carcinoma of the prostate: results of radical radiotherapy (1970–1985). Int J Radiat Oncol Biol Phys 26:203–210

Forman JD, Warmelink C, Devi S, Court W, Sharma R, Yudelev M, Porter AT, Maughan R (1995) Alternating conformal neutron and photon irradiation for locally advanced adenocarcinoma of the prostate. Am J Clin Oncol 18:231–238

Forman JD, Duclos M, Sharma R, Chuba P, Hart K, Yudelev M, Devi S, Court W, Shamsa F, Littrup P, Grignon D, Porter A, Maughan R (1996a) Conformal mixed neutron and photon irradiation in localized and locally advanced prostate cancer: preliminary estimates of the therapeutic ratio. Int J Radiat Oncol Biol Phys 35:259–266

Forman JD, Shamsa F, Maughan RL, Duclos M, Orton C (1996b) Comparison of hyperfractionated conformal photon with conformal mixed neutron/photon irradiation in locally advanced prostate cancer. Bull Cancer Radiother 83(s):101s–105s

Freiha FS, Bagshaw MA (1984) Carcinoma of the prostate: results of postirradiation biopsy. Prostate 5:19–25

Fuks Z, Leibel SA, Wallner KE, Begg CB, Fair WR, Anderson LL, Hilaris BS, Whitmore WF (1991) The effect of local control on metastatic dissemination in carcinoma of the prostate: Longterm results in patients treated with ^{125}I implantation. Int J Radiat Oncol Biol Phys 21:537–547

Fukunaga-Johnson N, Sandler HM, Mc Laughlin PW, Strawderman MS, Grijalva KH, Kish KE, Lichter AS (1997) Results of 3D conformal radiotherapy in the treatment of localized prostate cancer. Int J Radiat Oncol Biol Phys 38:311–316

Griffin TW, Laramore GE (1995) Neutron radiation and prostate cancer. Int J Radiat Oncol Biol Phys 33:231–232

Griffin TW, Pajak TF, Laramore GE, Davis LW (1986) Analysis of radiotherapy treatment complications. Bull Cancer (Paris) 73:582–586

Hanks GE, Leibel S, Krall JM, Kramer S (1985) Dose response observations for local control of adenocarcinoma of the prostate. Int J Radiat Oncol Biol Phys 11:153–157

Hanks GE, Martz KL, Diamond JJ (1988) The effect of dose on local control of prostate cancer. Int J Radiat Oncol Biol Phys 15:1299–1305

Hanks GE, Diamond JJ, Krall JM, Martz KL, Kramer S (1987) A ten-year follow-up of 682 patients treated for prostate cancer with radiation therapy in the United States. Int J Radiat Oncol Biol Phys 13:499–505

Hanks GE, Lee WR, Hanlon AL, Hunt M, Kaplan E, Epstein BE, Movsas B, Schultheiss TE (1996) Conformal technique dose escalation for prostate cancer: biochemical evidence of improved cancer control with higher doses in patients with pretreatment prostate-specific antigen ≥10 ng/ml. Int J Radiat Oncol Biol Phys 35:861–868

Haraf DJ, Rubin SJ, Sweeney P, Kuchnir FT, Sutton HG, Chodak GW, Weichselbaum RR (1995) Photon neutron mixed-beam radiotherapy of locally advanced prostate cancer. Int J Radiat Oncol Biol Phys 33:3–14

Kuban DA, El-Mahdi AM, Schellhammer PF (1989) Prognosis in patient with local recurrence after definitive irradiation for prostatic carcinoma. Cancer 63:2421–2425

Laramore GE, Krall JM, Thomas FJ, Griffin TW, Maor MH, Hendrickson FR (1985) Fast neutron radiotherapy for locally advanced prostate cancer: results of an RTOG randomized study. Int J Radiat Oncol Biol Phys 11:1621–1627

Laramore GE, Krall JM, Thomas FJ, Russell KJ, Maor MH, Hendrickson FR, Martz KL, Griffin TW, Davis LW (1993) Fast neutron radiotherapy for locally advanced prostate cancer: final report of a radiation therapy oncology group randomized clinical trial. Am J Clin Oncol 16:164–167

Russell KJ, Laramore GE, Krieger JN, Wiens LW, Griffeth JT, Koh W-J, Griffin BR, Austin-Seymour MM, Griffin TW, Davis LW (1990) Transient and chronic neurological complications of fast neutron radiation for adenocarcinoma of the prostate. Radiother Oncol 18:257–265

Russell KJ, Caplan RJ, Laramore GE, Burnison CM, Maor MH, Taylor ME, Zink S, Davis LW, Griffin TW (1993) Photon versus fast neutron external beam radiotherapy in the treatment of locally advanced prostate cancer: results of a randomized prospective trial. Int J Radiat Oncol Biol Phys 28:47–54

Scardino PT, Frankel JM, Wheeler TM, Meacham RB, Hoffmann GS, Seale C, Wilbanks JH, Easley J, Carlton CE Jr. (1986) The prognostic significance of post-irradiation biopsy results in patients with prostatic cancer. J Urol 135:510–516

Schellhammer PF, El MAM, Higgins EM, Schultheiss TE, Ladaga LE, Babb TJ (1987) Prostate biopsy after definitive treatment by interstitial 125 iodine implant or external beam radiation therapy. J Urol 137:897–901

Stamey TA, Ferrari MK, Schmid HP (1993) The value of serial prostate specific antigen determinations 5 years after radiotherapy: steeply increasing values characterize 80% of patients. J Urol 150:1856–1859

Wingo PA, Tong T, Bolden S (1995) Cancer statistics, 1995. CA Cancer J Clin 45:8–30

Zagers GK, von Eschenback AC, Ayala AG, Schultheiss TE, Sherman NE (1991) The influence of local control on metastatic dissemination of prostate cancer treated by external beam megavoltage radiation therapy. Cancer 68:2370–2377

The Role of Fast Neutrons in the Treatment of Squamous Cell Carcinomas of the Head and Neck: The European Experience

T. Auberger[1] and W. Reuschel[2]

[1] Department of Radiotherapy and Radiooncology, Leopold-Franzens University, Anichstrasse 35, A-6020 Innsbruck, Austria
[2] Department of Radiotherapy and Radiooncology, Klinikum rechts der Isar, Technical University, Ismaningerstrasse 22, D-81675 Munich, Germany

Abstract

After the first European stet on neutron therapy of head and neck cancer, published by Catterall and Bewley, revealed significant advantages of neutrons over photons in 134 randomized patients, the subsequent phase II and phase III studies led to controversial results. Figures for local tumor control varied from 26–76% after 2 years to 19% after 10 years and for survival from 16–46% after 2 years to 14% after 10 years. Results of all studies were consistent in showing an increased incidence of severe late effects. Eligibility criteria with regard to the tumor stage, histology, and previous therapies of relapses varied from study to study. Also, physical treatment parameters, such as neutron energy, LET and OER, neutron and photon doses for one fraction and for the total radiotherapy were rather different. The benefit of neutron therapy in comparison with photons was low or non-existent when smaller tumor stages, such as primaries T2 or lymph nodes N1 were included into the trial. It increased when only T4 tumors, recurrences or tumors persisting after conventional radiotherapy were treated and when the percentage of fixed lymph nodes increased. There was only one European study in which patients suffering from lymph node metastases of the neck had, in contrast to American data (Griffin et al. 1978), a worse prognosis after neutrons than after photon therapy (Duncan et al. 1987b). In our own phase II study on reactor neutron therapy of 100 patients suffering from relapses and persistent tumors, not yet published, 3-year survival of 21% and 3-year local control of 32% were obtained. Although this was not a randomized study, during this period all patients with equivalent tumors treated by photon therapy alone died within 2 years. In conclusion, there is no general indication for neutron therapy in squamous cell carcinoma of the head and neck, but there is a benefit of neutrons in palliative treatment of recurrences, necrotic and hypoxic primaries and lymph nodes.

Recent Results in Cancer Research, Vol. 150

Introduction

Since many clinical studies have shown that fast neutron therapy is the most effective treatment for well-differentiated and slowly proliferating salivary gland tumors (Catterall and Errington 1987; Kaul et al. 1981; Krüll et al. 1990; Kurup et al. 1982; Laramore 1987; Tsunemoto et al. 1989; Wambersie 1990), the role of fast neutrons in radiotherapy of advanced squamous cell carcinomas of the head and neck has been controversial in the international literature (Griffin et al. 1984, 1989; Kurup et al. 1982; Maor et al. 1986; Stannard et al. 1995; Tsunemoto et al. 1982; Wambersie 1990).

Randomized Studies

The first European data on fast neutron therapy of head and neck cancer from Hammersmith Hospital, London, in the 1970s were encouraging (Catterall and Bewley 1977; Catterell and Errington; Catterall et al. 1987). In a multicenter study from 1971 to 1975, reported by Catterall and Bewley (1977), 134 patients suffering from advanced squamous cell carcinoma were randomized into a fast neutron arm treated with a total dose of 15–17 Gy and a photon arm of 45–70 Gy (Table 1). The neutron therapy led to a local tumor control rate of 76% after 2 years, in comparison with 19% after photons. The overall survival rates after 2 years were 28% in the neutron arm and 15% in the photon arm, respectively. The rate of severe side effects such as skin and soft tissue necrosis increased from 3% after photons to 14% after neutron therapy, but, with regard to the better survival, the complication rate seemed to be tolerable. The study came in for much criticism because only a small part of the patients in the photon arm were treated at Hammersmith Hospital and therefore in part the photon therapy was not performed according to modern standard treatment techniques. However we have to respect that also nowadays we cannot reach much better 2-year local control and survival rates than 20% in advanced head and neck cancer (T4, N2-3, recurrences).

A subsequent randomized study was performed in Edinburgh from 1977 to 1982 in 165 patients with head and neck primaries and was reported by Duncan et al. (1987a). Despite salivary gland tumors also being included in this trial, no advantage of neutron therapy over conventional radiotherapy could be observed with regard to local control and 5-year survival. The results were consistent in associating an increased incidence of severe late radiation morbidity and of intercurrent deaths. However, in the Edinburgh trial there was an advantage to the photon-treated patients in terms of overall mortality, whereas at Hammersmith the advantage was to those treated with neutrons. In 1986 the Medical Research Council Neutron Therapy Working Group of Great Britain compared the first 120 patients entered into the Edinburgh trial with 95 patients selected from the Hammersmith trial. The important aspects in which the two investigations had differed included the

Table 1. Randomized and not randomized European studies on neutron therapy or mixed photon-neutron therapy of advanced squamous cell carcinoma of the head and neck

Institute	Authors	Trial	Tumors	Patients (*n*)		Total dose (Gy)	
				Neutrons	Photons	Neutrons	Photons
Hammersmith 1982	Catterall and Errington (1987)	Phase III	Primaries (T3, T4), relapses	70	63	15–17	45-69
Amsterdam 1986	Battermann and Mijnhofer (1986)	Phase II	Advanced primaries, no recurrences	65	–	18–21	–
Edinburgh 1984	Duncan et al. (1987 a)	Phase IIII	Primaries (including salivary gland tumors)	85	80	15–17	54–56
EORTC 1984 Update 1990	Duncan et al. (1984) Duncan (1994) MacDougall et al. (1990)	Phase III	Primaries T2,3 and N0,1 mainly	100	95	15–16,5	70(35f) or 54(20f)
Krakow 1988	Skolyszewski et al. (1988) Skolyszewski and Reinfuss (1985)	Phase II	Advanced primaries and recurrences	37	–	13	–
Obninsk 1991	Mardynsky et al. (1991)	Phase II	Only primaries of laryngeal carcinoma	78	–	2,5–7 n + 12–30 ph	
Hammersmith, Clatterbridge 1987	Catterall et la. (1987)	Phase II	Advanced tumors and relapses	28	–	15,6	
Munich 1996	Auberger et al. (complete results not published)	Phase II	Relapses and persistent tumors after radiotherapy	100	–	30 ph + 6 n 6 n	

stage of disease (at Hammersmith T3, T4 only and 63% of patients having fixed nodes, in Edinburgh T2 included and only 27% of patients having fixed nodes) and the radiation dosage (median neutron dose being 5% lower in Edinburgh than at Hammersmith while the median photon dose was 10% higher in Edinburgh).

From 1975 to 1981 in the Netherlands Cancer Institute in Amsterdam, 65 head and neck tumors (salivary gland tumors not included) were irradiated by a 14-MeV D+T neutron generator (Battermann and Breur 1981; Battermann and Mijnheer 1986). Only patients with extended tumors with a likelihood of less than 10% of cure by conventional photon therapy were chosen. The total physical dose including 10%–20% gamma contribution varied between 18 and 21 Gy (20 fractions within 4 weeks). As routine, cervical lymph nodes were first included within the treatment fields and if these nodes were involved the whole neck was irradiated. To reduce the dose to the spinal cord, fields were reduced after 12 Gy and an additional 10–20 Gy was administered by electrons to the area overlying the spinal cord.

Fifty-seven patients were analyzed. All patients died. The median survival was 12 months. In many cases the tumor regression was faster than seen after conventional photon irradiation. Complete remission (CR) at 2 months from start of irradiation was found in 38/57 patients, but persisting local control was achieved in only 26 patients. Severe complications were noticed in seven patients. Complications were myelopathy, brain damage related to the wide penumbra of the D+T beam and complications in the larynx. The results of the combined head and neck trial were not significantly different for both treatment arms. Although the remission rates at 2 months were better after neutron therapy than after photons, the survival was longer after conventional radiotherapy (29% vs. 15% at 2 years). Also, in contrast to others, the persisting control in positive nodes was better after photon irradiation (Duncan et al. 1987b).

Subsequently 20 patients treated with neutrons and 13 patients treated with photons were entered into a multicenter-EORTC study. Tumors of the base of skull (paranasal sinuses, nasopharynx carcinomas) and early stages of tumors in the tongue and larynx were excluded. The neutron dose was 18 Gy in 20 fractions within 4 weeks.

In 1984 Duncan et al. reported the results of all contributors of this European multicenter study on fast neutron therapy in patients with cancer of the oral cavity, oropharynx, hypopharynx and larynx. This randomly controlled EORTC trial compared neutron therapy of 15–16.5 Gy (dependent on field size, beam quality and localization) in 20 fractions over a period of 4 weeks and photon therapy of 70 Gy in 35 fractions or 54 Gy in 20 fractions. The participating centers were the Antoni van Leeuwenhock Hospital in Amsterdam, the Department of Clinical Oncology, Edinburgh, and the University Clinic, Essen. Altogether the three centers randomized 100 patients into the neutron arm and 95 patients into the photon arm. In order to employ equal effective doses despite different beam qualities (15-MeV D+T generator in Amsterdam, compact cyclotrons with a neutron beam of 6 MeV on average in

Edinburgh and Essen), in the protocol the specification of neutron dose was especially taken into account. Tumors previously treated by radiotherapy, chemotherapy or surgery were excluded from this trial. The majority of tumors were stage T 2 or T 3. Nearby 50% of the lymph node metastases were stage N 1 a or smaller. Tumors with high probability of local control after photon therapy, such as T 1 or T 2 laryngeal carcinoma and T 1 or T 2 carcinoma of the oral cavity, were excluded. Tumors of the nasopharynx and paranasal sinuses were also excluded because of the low tolerance of surrounding critical organs. The study could not reveal any significant benefit of neutrons with regard to the outcome and to severe side effects. The rates of complete tumor remission were similar after neutrons (70%) and photons (66.3%) and the numbers of recurrences were also similar (37.1% in the neutron group, 39.7% in the photon group). The overall survival after 3 years was 33% v. 39% and the local tumor control 34% v. 39%. There was a higher number of overall fatal complications in the group treated with neutrons, 6% compared to 2% in the photon group, but this difference did not reach statistical significance. Seventeen patients in the neutron-treated group and ten patients in the group treated by photon irradiation developed mucosal necrosis. The total serious complications in the mucosa were similar after neutrons and photons.

An update of Duncans' results published by MacDougall et al. in 1990 focused on the outcome in the 27 patients who developed necrosis after radiotherapy. Nine patients remained alive. In five cases necrosis had been healed by conservative treatment, in four cases by surgery. Of the 18 patients who had died, the causes were intercurrent deaths in two patients of each group and recurrences in four patients after neutrons and three patients after photons, respectively. Seven patients died because of the radiation-induced necrosis in the neutron group but no patients died in the photon group.

Non-randomized Studies

Between 1978 and 1989, 164 patients suffering from advanced head and neck cancer were treated with fast neutron therapy at the Center of Oncology in Krakow, Poland. Results were evaluated retrospectively and reported by Skolyszewski and Reinfuss (1985) and Skolyszewski et al. (1988) and updated by Sas-Korczynska and Skolyszewski (1996). Three schedules of dose fractionation were used: 114 patients were irradiated with a dose of 12–13.2 Gy in 20 fractions. Seventeen patients received 10–12 Gy in 10 fractions and 33 patients received 10–13 Gy in 5 fractions. Complete remission after therapy was observed in 35.4% of the patients but only 26% of the patients survived 2 years without disease. There were no significant differences in survival rates between patients treated with the three different dose fractionation schedules. This supports the thesis that in high-linear energy transfer (LET) irradiation the single dose does not substantially influence the treatment results. Twenty-four patients developed severe late complications after neutron therapy (mucosal necrosis in 8 patients, mandible necrosis in 6 patients, severe laryngeal edema in 9 patients, cervical myelopathy in 1 patient.

At the reactor neutron source of the Russian Academy of Medical Science in Obninsk near Moskow, Mardynsky et al. (1991) treated 87 laryngeal carcinomas by mixed-beam radiotherapy consisting of 2.5–7 Gy reactor neutrons and 12–30 Gy photons. They reported a local control rate of 70% after 2 years, but they were unable to follow-up more than 27 patients within this time. The complications were perichondritis in two patients. In some European neutron therapy facilities (Hamburg, Berlin, Heidelberg) small groups of patients with advanced head and neck cancer were treated by neutron irradiation for palliation, but no statistical data were evaluated.

Neutrons for Palliative Therapy of Recurrences

Only few data exist for palliative neutron therapy in head and neck cancer (Auberger et al. 1992; Catterall et al. 1987; Saroja et al. 1988; Skolyszewski and Korzeniowski 1988). In non-randomized common series at Hammersmith and Clatterbridge, reported by Catterall et al. (1987), 28 patients suffering from extremely advanced tumors and recurrences after previous radiotherapy were treated with neutrons for palliation. A total dose of 15.6 Gy was delivered in 12 fractions over 28 days. Of these advanced tumors, 82% completely regressed for more than 1 year. The complication rate in the previously irradiated patients was 46%. In comparison, for patients who had not undergone previous radiotherapy the complication rate was 27%. Taking into account the high local control rate and the poor prognosis of advanced uncontrolled tumors, the complication rate can be regarded as acceptable. The incidence of bone necrosis was even lower. Not uncommonly a healing of bone defects, which had developed because of tumor progress, could be observed.

Most patients treated by neutrons for palliation of advanced head and neck cancer were investigated in the RENT (reactor neutron therapy) facility of the research reactor at the Technical University of Munich and in the Radiooncological Department of the Klinikum rechts der Isar, Munich. One hundred patients suffering from advanced squamous cell carcinoma of the head and neck were either treated by mixed-beam radiotherapy or by neutron therapy alone (Auberger et al. 1992). The RENT facility at Munich delivers a beam consisting of 75% fission neutrons (mean energy: 1.8–2 MeV, range of the spectrum: 1–7 MeV) and of 25% photons (mean energy equivalent to a cobalt source). An advantage of the very low energy of fission neutrons is the high LET and therefore also a very high relative biological effectiveness (RBE) (RBE for normal soft tissues >3 for 5 fractions) and the lowest oxygen enhancement ratio (OER) of all working neutron therapy facilities (OER $\sim$1.3). A disadvantage of fission neutrons is the low penetration depth in tissue (50% isodose in 5 cm depth of tissue). Because of the high efficiency on hypoxic tumors (in comparison with normal tissues) we focused on previously irradiated recurrences and lymph node metastases.

Of the 100 patients suffering from head and neck cancer, 56 had a local recurrence and 44 patients had lymph node metastases. Most had previously been treated by radiation therapy (70–120 Gy)± surgery or chemotherapy. For the mixed-beam therapy two different protocols were performed: (1) in persistent primaries and lymph node metastases 4–8 weeks after photon radiotherapy of 70 Gy (group A – 27 patients) additionally a neutron boost of 1–3 fractions of 2 Gy (1 fraction/week) was applied. (2) Recurrences and lymph node metastases after a previous radiotherapy±surgery±chemotherapy more than half a year previously (group B: 48 patients) were treated with mixed-beam therapy of 30 Gy photons or electrons (5 fractions of 2 Gy/week) and 6 Gy neutrons (1 fraction of 2 Gy/week). (3) Recurrences and lymph node metastases after more than one previous radiotherapy session (group C: 25 patients) and metastases, which were not operable or showed no local response to chemotherapy, were treated by neutrons alone up to a total dose of 6–10 Gy (1–2 fractions of 2 Gy/week).

Complete remissions (CRS) were 74% in group A and 65% in group B. These are good results taking into account the poor prognosis of repeated photon therapy in persistent and recurrent tumors. In group C, with regard to the small total dose and to the absence of any other treatment, a CR of 20% and a partial remission (PR) of another 45% also revealed a remarkable palliative effect. Local tumor control in the total patient group after 2 and 3 years was 36% and 32%, respectively. The actuarial overall survival was 21% after 2 years and 14% after 3 years. In correlation with CR rates the best survival was observed in group A. Survival rates after 2, 3 and 5 years were 45%, 38% and 30%, respectively (Table 2). The local tumor control after 3 years was 55% in this group.

The relatively poor survival was influenced by the high number of intercurrent deaths (30%). Subsequently, 21% of the patients died from metastases (patients who already had metastases at the beginning of neutron therapy were included in the evaluation) and only 46% from locoregional tumor progress.

With regard to the previous multiple therapies, the grade of late reactions was moderate and the rate of severe complications was small. One-third of the patients showed no major reactions. The main complications were lymph edema ($n=16$) and tumor necrosis ($n=17$). In 11 out of the 17 patients tumor necrosis resulted in spontaneous bleeding, which did not lead to death in any patient. However, two patients also presented a hemorrhage without microscopic tumors in the irradiated field 1–3 years after neutron therapy.

In conclusion, these results shwo there is a benefit of reactor neutron therapy in treatment of persistent tumors after the end of conventional photon therapy (boost) and in palliative therapy of recurrences more than half a year after previous radiotherapy. Survival of patients is determined in more than 50% by metastases and intercurrent deaths because of the high general morbidity of this patient group. It is an important therapeutic gain to achieve local control in these patients until death. Reactor neutron therapy seems to be a method which can help to achieve this gain with relatively few side effects and with a remarkable improvement of the quality of life.

Table 2. Results of European studies on neutron therapy and mixed photon-neutron therapy in advanced squamous cell carcinomas of the head and neck

Institute	CR (%)		Survival (%)			Local control l(%)			Complications (%)	
	Neutrons	Photons	Years	Neutrons	Photons	Years	Neutrons	Photons	Neutrons	Photons
Hammersmith 1982	77	43	2	28	15	2	76	19	14	3
Edinburgh 1984	71	78	5	24	34	5	44	45	–	–
Amsterdam 1986	66	–	2	46		2	46	–	12	–
EORTC 1984	70	66	3	33	39	3	34	39	22	25
Update 1990		–	10	14	30	10	19	30	21	11
Krakow 1988	54	–	2	16	–	2	26	–	7	–
Obninsk 1991	no data					2	70	–	7	–
Hammersmith, Clatterbridge 1987	82	–	–	–	–	1	82	–	46	–
Munich	65	–	3	21	–	3	32	–	14	–

CR, complete remission; *Complications*, skin and soft tissue necrosis.

Similar results were shown repeatedly in oral presentations by Errington, i.e., during the EHTG meeting in 1995 in Orléans, France.

Conclusion

Most randomized studies on neutron therapy for primaries of head and neck squamous cell carcinomas have concluded that in general there is no benefit of neutrons over conventional photon or electron therapy. With request to regard this question of primaries European studies are continuing along with the multicenter studies in the United States (Griffin et al. 1984, 1989; Kurup et al. 1982; Maor et al. 1986) and studies in South Africa (Stannard et al. 1995) and Japan (Tsunemoto et al. 1982, 1989).

However, in many studies the recruitment of patients has not been clearly defined with respect to the special radiobiological effects of neutrons. Tumors of different stages and sizes in some studies (even T2) were included in the protocols without regard to the different radiosensitivity of individual tumors. Special biological conditions (large tumor masses with necrotic areas, reduced vascularization because of previous therapies) define subgroups of head and neck cancer which can have a benefit in neutron therapy. Also the American multicenter studies on neutron therapy of head and neck cancer reported by Griffin et al. (1978, 1984) resulted in a better outcome for subgroups such as advanced lymph node metastases.

Especially in head and neck cancer local progress of the primary or of the recurrence is combined with a tremendous loss of quality of life and it takes time before this local progress leads to death of the patient. So we must try to achieve the best palliation we can in this cases.

The most important argument against neutron therapy is the higher number of severe side effects of neutrons in comparison with photons and electrons. The high rate of intercurrent deaths and the many metastases at the beginning of treatment emphasize the fact that even after a most radical and effective treatment the overall survival will be rather low. Therefore the reduction in quality of life by these radical treatment modalities must be especially considered. On the other hand, it has to be taken into account that tumor progression causes many severe side effects. The number of side effects can be reduced significantly by improving the physical selectivity of the beam. Studies of conformal neutron therapy of prostatic cancer could confirm this fact. An even more precise treatment modality may be light ion therapy, which combines the advantages of the physical selectivity, as known from protons, and the high RBE of neutrons. The considerable improvement of precise dose distribution of light ions could also lead to renewed evaluations of head and neck cancer therapy.

References

Auberger T, Reuschel W, Mayr M, Kneschaurek P, Lukas P, Clasen B, Breit A (1992) Mixed-beam photon-neutron therapy in recurrences and nodal metastases of head and neck cancer. In: Breit A (ed) Advanced radiation therapy tumor response monitoring and treatment planning, Springer, Berlin Heidelberg New York, pp 803–810

Battermann JJ, Breur K (1981) Results of fast neutron teletherapy for locally advanced head and neck tumors. Int J Radiat Oncol Biol Phys 7:1045–1050

Battermann JJ, Mijnheer BJ (1986) The Amsterdam fast neutron therapy project: a final report. Int J Radiat Oncol Biol Phys 12:2093–2099

Catterall M, Bewley DK (1977) Second report on results of a randomized clinical trial of fast neutrons compared with X or gamma rays in treatment of advanced tumours of head and neck. Br Med J 1:1642

Catterall M, Errington RD (1987) The implications of improved treatment of malignant salivary gland tumors by fast neutron radiotherapy. Int J Radiat Oncol Biol Phys 13:1313–1318

Catterall M, Errington RD, Bewley D (1987) A comparison of clinical and laboratory data on neutron therapy for locally advanced tumors. Int J Radiat Oncol Biol Phys 13:1783–1791

Duncan W (1994) An evaluation of the results of neutron therapy trials. Acta Oncol 33:299

Duncan W, Arnott SJ, Battermann JJ, Orr JA, Schmitt G, Kerr GR (1984) Fast neutrons in the treatment of head and neck cancers: the results of a multi-centre randomly controlled trial. Radiother Oncol 2:293–300

Duncan W, Orr JA, Arnott SJ (1987a) Fast neutron therapy for squamous cell carcinoma in the head and neck region: results fo a randomized trial. Int J Radiat Oncol Biol Phys 13:171–178

Duncan W, Orr JA, Arnott SJ, Jack WJL, Kerr GR (1987b) An evaluation of fast neutron irradiation in the treatment of squamous cell carcinoma in cervical lymph nodes. Int J Radiat Oncol Biol Phys 13:1793–1796

Griffin TW, Laramore GE, Parker RG, Gerdes AJ, Hebard DW, Blasko JC, Groundine M (1978) An evaluation of fast neutron beam teletherapy of metastatic cervical adenopathy from squamous cell carcinomas of the head and neck. Cancer 42:2517–2520

Griffin TW, Davis R, Hendrickson FR, Maor MH, Laramore GE (1984) Fast neutron radiation therapy for unresectable squamous cell carcinomas of the head and neck: the results of a randomized RTOG study. Int J Radiat Oncol Biol Phys 10:2217–2221

Griffin TW, Pajak TF, Maor MH, Laramore GE, Hendrickson FR, Parker RG, Thomas FJ, Davis LW (1989) Mixed neutron/photon irradiation of unresectable squamous cell carcinomas of the head and neck: the final report of a randomized cooperative trial. Int J Radiat Oncol Biol Phys 17:959–965

Kaul R, Hendrickson F, Cohen L (1981) Fast neutrons in the treatment of salivary gland tumors. Int J Radiat Oncol Biol Phys 7:1667–1671

Krüll A, Schwarz R, Heyer D, Brockamnn WP, Junker A, Schmidt R, Hübener KH (1990) Results of fast neutron therapy of adenoidcystic carcinomas of the head and neck at the neutron facility Hamburg-Eppendorf. Strahlenther Onkol 166(1):107–110

Kurup PD, Hendrickson F, Cohen L (1982) Radiation therapy utilizing fast neutrons for head and neck cancer – the Fermilab experience. Int J Radiat Oncol Biol Phys 8 Suppl 1:103–104

Laramore GE (1987) Fast neutron radiotherapy for inoperable salivary gland tumors: is it the treatment of choice? Int J Radiat Oncol Biol Phys 13:1421–1423

Medical Research Council Neutron Therapy Working Group (1986) A comparative review of the Hammersmith (1971–75) and Edinburgh (1977–82) neutron therapy trials of certain cancers of the oral cavity, oropharynx, larynx and hypopharynx. Br J Radiol 59:429–440

MacDougall RH, Orr JA, Kerr GR, Duncan W (1990) Fast neutron treatment for sqaumous cell carcinoma of the head and neck: final report of Edinburgh randomised trial. Br Med J 301:1241–1242

Maor MH, Schoenfeld DA, Hendrickson FR, Davis LW, Laramore GE, Thomas FJ, Pajak TF (1986) Evaluation of a neutron boost in head and neck cancer. Am J Clin Oncol 9:61–66
Mardynsky YS, Sysoyer AS, Andreyev VG, Gulidov IA (1991) Preliminary rsults of clinical application of reactor fast neutrons in radiation and combined therapy of patients with laryngeal carcinoma. Strahlenther Oncol 167:169–171
Sas-Korczynska B, Skolyszewski J (1996) Acute and late reactions of fast neutron therapy in patients with head and neck cancer – the Krakow experience. In: Current status and future directions in particle therapy. Proceedings of the XXIV PTCOG Meeting 24–26 April
Saroja KR, Hendrickson FR, Cohen L, Mansell J, Lennox A (1988) Re-irradiation of locally recurrent tumors with fast neutrons. Int J Radiat Oncol Biol Phys 15:115–121
Skolyszewski J, Korzeniowski S (1988) Re-irradiation of recurrent head and neck cancer with fast neutrons. Br J Radiol 61:527–528
Skolyszewski J, Reinfuss M (1985) Results of fast neutron therapy for advanced carcinomas of the oral cavity and oropharynx. Radiobiol Radiother 26:85–89
Skolyszewski J, Korzeniowski S, Huczkowski J (1988) Results of fast neutron therapy in advanced head and neck cancer. Br J Radiol 61:301–304
Stannard C, Vernimmen F, Jones D, Wilson J, van Wijk L, Brennan S, Schreuder N, Symons J, Levin V, Mills E, Alberts A, Werner D, Smit B, Schmitt G (1995) Neutron therapy program at the national acelerator centre, South Africa – preliminary results: Radiat Oncol Invest 2:245–255
Tsunemoto H, Arai T, Morita S, Ishikava T, Aoki Y, Takada N, Kamata S (1982) Japanese experience with clinical trials of fast neutrons. Int J Radiat Oncol Biol Phys 8:2169–2172
Tsunemoto H, Morita S, Satoh S, Iino Y, Yoo Y (1989) Present status of fast neutron therapy in Asian countries. Strahlenther Onkol 165:330–336
Wambersie A (1990) Fast neutron therapy at the end of 1988 – a survey of the clinical data. Strahlenther Onkol 166:52–60

Sacrococcygeal Chordomas: Potential Role of High-LET Therapy

N. Breteau[1], M. Demasure[1], J. Lescrainier[1], R. Sabbattier[1] and P. Michenet[2]

[1] CHR Service D'Oncologie et de Radiothérapy, BP 6709, F-45067 Orléans Cedex 2, France
[2] CHR Service d'Anatomopathologie, BP 6709, F-45067 Orléans, France

Abstract

Chordomas are rare malignant neoplasms representing less than 3% of all primary bone tumors. They usually have a benign histological appearance and a slow growth rate. Their locoregional progression nearly always causes suffering and eventually death. Efforts to increase local control have included aggressive surgery, radiation therapy or combined approaches but locally free survival rates remain relatively low and have not exceeded 30% at 5 years. Sacrococcygeal chordomas, which represent approximately 50% of the localizations, have been investigated less frequently than those at the base of the skull or cervical region with primary or post-operative radiation therapy. The disappointing results with photon therapy in a multimodality approach and the good results reported by Schoenthaler with charged particles in the Lawrence Berkeley laboratory led us to propose fast neutron therapy in the management of inoperable or recurrent sacrococcygeal chordomas.

Preliminary results obtained from only 12 patients indicate that high linear energy transfer (LET) therapy seems to be a good alternative for radical treatment of chordomas in the case of microscopic or macroscopic residual tumor.

Introduction

Chordoma is a rare neoplasm arising from notochordal remnants in the midline of the neural axis (Pena et al. 1970; Spjut and Luce 1964) and involving the adjacent bone. These tumors represents less than 3% of all primary bone tumors (Wright 1967). Both ends of the spine are the most common sites. About 50% arise in the sacrococcygeal area, and 35% arise intracranially, where they typically involve the clivus. The remaining 15% occur in the midline along the path of the notochord, primarily involving the cervical vertebrae (Cummings et al. 1983; Riche et al. 1985; Tewfik et al. 1977). Chordomas have been reported in every age group but most commonly affect patients in the 6th or 7th decades (Kaiser et al. 1984). For children and young

Recent Results in Cancer Research, Vol. 150

adults, the prognosis and long-term survival rate appear to be better than for older patients (Wold and Laws 1983). These tumors occur two or three times more often in males than in females but no risk factors have been identified as yet.

The lethality of these tumors rests on their critical location aggressive local behavior and their extremely high local recurrence rate. Efforts to increase local control have included aggressive surgery, radiation therapy or combined approaches but the results have been disappointing. It seems that as far as sacrococcygeal location is concerned a therapeutic window can be identified for fast neutron therapy.

Clinical Presentation and Diagnosis

Chordomas are slow-growing lesions. They are locally invasive, destroy bone and infiltrate soft tissues. The most common complaint of patients suffering from sacrococcygeal tumors is a dull pain, with constipation as an occasional symptom and bladder sensory loss and lower extremity neurologic deficits as late complaints. Rectal examination characteristically reveals a large presacral mass. Computed tomography (CT) and magnetic resonance imaging (MRI) are essential for accurate evaluation. Most patients have significant bone destruction and some may have calcifications in the tumor. A CT scan with contrast enhancement is required to reveal the soft tissue component. If CT and MRI are equivalent in terms of revealing the presence and the site of the tumors MRI is superior to CT for delineating the extent of the tumor, which allows better treatment planning.

A transrectal biopsy should not be performed because of the potential contamination. A small midline posterior incision or trocar biopsy is recommended: chordoma is a soft, lobulated tumor that may have areas of hemorrhage, cystic changes or calcification. It is frequently encapsulated but may be nonencapsulated or pseudocapsulated. Histologically, the physaliferous cells are pathognomonic. Large areas of syncytial strands of cells lying in a mass of mucus are typically present. There is usually a small number of mitotic cells (Erlandson et al. 1968; O'Neil et al. 1985).

Chordoma is one of the most relentlessly destructive tumors encountered in clinical practice. Locoregional progression nearly always causes suffering and eventually death (Dahlin and MacCarty 1952; Higinbotham et al. 1967; Pearlman and Friedman 1970). The incidence of metastases is higher than previously believed, found in 15%–25% of cases (Higinbotham et al. 1967; Saunders et al. 1981). The most common sites of distant metastases are the lungs, followed by the liver and the bones. Lymphatic spread is uncommon (Wang and James 1968).

Classical Treatment

Surgery

The first surgical procedure is well known to have the best probability of cure (Gray et al. 1975), but even for small lesions where radical sugery seems to be the most effective, recurrence rates are still high, ranging from 50% to 100% (Cummings et al. 1983, Gray et al. 1975; O'Neil et al. 1985; Saxton 1981). For large tumors, total removal is usually impossible and partial excision is almost universally followed by local failure (Cummings et al. 1983; Fuller and Bloom 1988; Gray et al. 1975, Reddy et al. 1981).

Radiation Therapy

Because local recurrence is common with chordoma, radiation therapy is an integral treatment modality particularly for tumors of the base of the skull and the spheno-occipital region. In that location, results of conventional radiation therapy even with high doses (from 50 to 70 Gy) (Amendola et al. 1986; Riche et al. 1985) used postoperatively have been disappointing, with an overall 5-year disease-free survival rate of less than 10%–20% (Amendola et al. 1986; Higinbotham et al. 1967; Saxton 1981). More recently several investigators have advocated the use of precise heavy charged particle irradiation (Austin-Seymour et al. 1986; Raffel et al. 1985) with proton beams at a medium tumor dose of 69 Gy eg. Cobalt gamma (GyEG). This has achieved an actuarial 5-year disease free survival of 76% and a local control rate of 82%.

Sacrococcygeal chordoma has been less frequently investigated with combined approaches. This situation probably results from the apparent possibility of cure by radical surgery alone for this slow-growing tumor and most of the data in the literature concern radiation therapy proposed after several unsatisfactory operations or as a palliative.

Nevertheless, several reports on the use of primary or postoperative photon therapy have demonstrated a radiation dosage response and good palliative results (Amendola et al. 1986; Cummings et al. 1983; Higinbotham et al. 1967; O'Neil et al. 1985; Pearlman and Friedman 1970; Reddy et al. 1981; Saxton 1981). However, locally free survival rates remain relatively low and exceed 30% at 5 years.

Clinical Material

Between 1981 and 1994, 13 patients with inoperable or recurrent pelvic chordomas were referred to the neutron therapy department. Two patients presented with a primary tumor and 11 with relapse. Three were referred for palliative treatment after previous surgery and radiation therapy (50 Gy or more). One of these three patients decided to discontinue the treatment after three sessions because of the high risk of complications.

Twelve patients were suitable for evaluation. As has already been noted by other authors (O'Neil et al. 1985; Schmitt and Schnabel 1984), we found a predominance of males with a ratio 9:3 males:females. There was a wide range of ages among our 12 patients, ranging from 8 to 80 years, with a mean of 62 years.

Two patients had very large tumors occupying the whole pelvis and were sent for therapy after surgical biopsy. The others were referred to the neutron therapy unit with inoperable relapses after one to five prior surgical procedures (mean 1.9). Two patients had relapses in the irradiated area. In accordance with the clinical progression of these slow-growing tumors, ten patients had very long histories before being referred to Orléans for neutron therapy, with an average delay of 45 months from the time of diagnosis (median 46 months). However, the delay between the last surgical operation and neutron therapy was 13 months (median). When studying the results and particularly the local control rate, one must consider this delay rather than the survival based on diagnosis.

All the patients were referred to neutron therapy with gross inoperable disease causing great pain. Ten out of 12 patients had a relatively complete cauda equina syndrome and one patient had intestinal obstruction syndrome which required colostomy.

MRI has been the gold standard for radiological examination in our series, allowing a precise definition of target volume particularly for bone extension. For technical reasons, patients were treated with a CT scan based dosimetry but taking into account the findings of the MRI examination.

Treatment Techniques

Tumors ranged from 6 to 36 cm in diameter (average 10 cm). According to the target volume or the purpose of treatment, neutrons were used either alone or as a boost. Indeed, taking into account the facility (fixed vertical beam) and the relationship between target volume and complication (Schmitt and Schnabel 1984), neutrons alone were used for tumors less than 10 cm in diameter. Seven patients with a mean tumor diameter of 16.8 cm received 40 Gy X-rays prior to the neutron boost. The neutron equivalent doses ranged from 15 GyEC for tumors more than 22 cm in diameter to 25 GyEC for a total dose of 55 to 65 Gy eg. delivered in conventional fractionation (10 Gy, 4 fractions, 1 week). Five patients had smaller tumors with a mean diameter of 8 cm, and were treated with neutrons alone. Two of them, who had been previously irradiated, received 10 Gy $n+\gamma$ in 12 fractions over 21 days. Three were treated with curative intent and received 17.6 Gy $n+\gamma$ in 16 fractions over 28 days (Table 1). CT scans were performed with the patient in the treatment position using an immobilization device. We generally used a

Table 1. Treatment procedure

	Target volume (mean diameter)	Nb
40 X-rays + 15 → 25 N eq GY	16.8 cm	7
Curative treatment		3
17.6 N Gy 16 f 28 d	8 cm	
Palliative treatment		
10 N GY 12 f 21 d		2

f, fraction; d, days

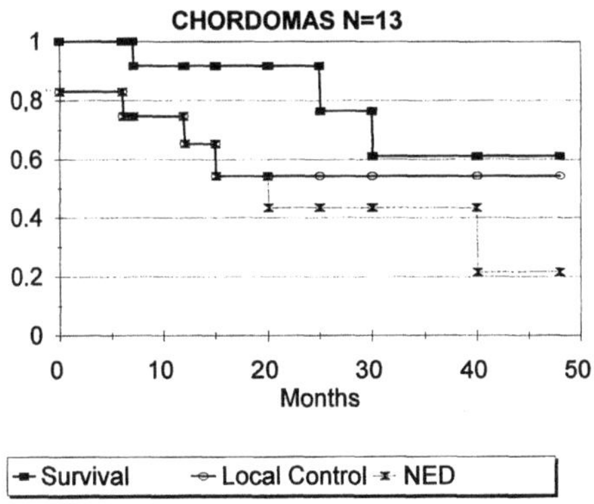

Fig. 1. Survival rate and local control probabilities

three-field technique with blocks and wedged beams, but in two cases due to the tumor volume we used a four-orthogonal field technique.

Results

According to the Kaplan-Meier method, crude survival at 4 years is 61% with a probability of local control of 54% (Fig. 1). The four patients treated with palliative intent (10 Gy $n+\gamma$ or 55 GyEC total dose) had a mean follow-up of 15 months (7–22 months). One patient died after 7 months due to tumor progression; two had pain and neurological relapse at 12 and 15 months, respectively after therapy. Seven of the eight patients treated with a curative dose either with neutrons at 17.6 Gy $n+\gamma$ or a neutron boost with a total dose equal to or greater than 60 GyEC are locally free of disease at the time of evaluation. The remaining patient died, locally free of disease, 30 months after treatment and after 24 months of metastatic progression. Two patients (out of the 12) developed metastases. One, with metastases in the ischiorectal fossa, was cured surgically. None or minimal acute side effects were observed for ten patients and moist desquamation developed for two patients treated with neutrons alone. Three patients with a large target volume developed diarrhea.

Complications occurred after 12 months and at the time of evaluation no grade 3 or 4 complications were observed. A girl, 8 years old at the time of treatment, is still alive and in good shape, more than 8 years later, with normal ovarian function; demonstrating that pelvic deformation is perfectly compatible with normal life (Table 2).

Discussion and Conclusion

These results have been compared with previous data (Table 3). In 1985, the Western General Hospital in Edinburgh published details of 34 cases of sacral chordoma treated either by surgery alone or followed by X-ray therapy with doses ranging from 40 to 50 Gy.

At 4 years, the survival probability was better for the photon group at 53% vs. 30%, demonstrating the beneficial effect of radiotherapy. Unfortunately the authors gave no information concerning local control or disease-free survival and only 8 out of 34 patients were alive with follow-up from 1 to 20 years with the absence of a plateau on the survival curves.

In 1992, Schoenthaler et al. reported 14 cases of sacral chordomas treated post operatively with charged particles of helium and neon between 1977 and 1989 at the Lawrence Berkeley Laboratory. In this series, only four patients had undergone total excision. The 5-year Kaplan-Meier survival was 85%, with a probability of local control of 55%. This study shows a trend to-

Table 2. Side effects

Nor or minimal acute reaction	10
Moist desquamation (neutrons alone)	2
Diarrheas (large target volume)	3
Late side effects (12 months) G0–2	12

G, grade

Table 3. Literature data

O'Neil et al (1985) Western General Hospital (Edinburgh, UK)	*n*=34
Sacral total excision + X-ray	53% 5-year survival probability
Surgery alone	30% 5-year survival probability
Schoenthaler et al. (1992) Lawrence Berkeley Laboratory (USA)	*n*=14
Sacral total excision + charged particle (62% with neon; 34% with helium)	55% 5-year local control probability
Higinbotham et al. (1967) Memorial Hospital (Boston, USA)	*n*=18
Macroscopic residual or inoperable tumors X-rays alone	49% objective response
Breteau et al. (1996) CHRO (Orléans, France)	*n*=13
Inoperable tumor or inoperable relapse	
Neutron or mixed	53% 3-year local control probability

ward improved local control following total resection, with 5-year local control rates of 75% and 50% following total and subtotal resection respectively. For high LET the 5-year local control probability was 62% with neon and 34% with helium.

We found only two consistent studies concerning exclusive radiotherapy of gross residual or inoperable sacrococcygeal chordomas. Higinbotham et al. in 1967 published details of a series 18 patients treated in The Memorial Hospital with macroscopic residual or inoperable tumors: 13 received symptomatic relief, 8 of these (49%) showed an "objective response of the tumor" (sic) and 5 patients did not show improvement.

In 1988, Fuller and Bloom reported results from 25 patients treated by radical radiotherapy after partial excision or biopsy: the overall actuarial 5-year survival and disease-free survival rates were 44% and 33%, respectively. In that work, the authors identified the concept "progression free interval," which seems adapted to chordomas. Of the nine patients with sacral disease, 52% were 5-year actuarial progression free (two persistent local control, three with no evidence of progression at 46, 90 and 113 months after radiotherapy).

In our series, no patient had surgical removal before neutron therapy, principally because of the tumor size and local involvement, which could have led to severe neurological complications. However, our results and those reported by Schoenthaler et al. (1992), showing a probability of local control of 54% and 62%, respectively, might indicate that using high LET for inoperable or macroscopic residual tumor is a good alternative for chordomas. Spinal cord or clivus locations are better treated with charged particles for obvious reason (Schoenthaler et al. 1992; Suit et al. 1982), but sacrococcygeal chordomas seem a good indication for fast-neutron therapy.

References

Amendola BE, Amendola MA, Oliver E, McClatchey KD (1986) Chordoma role of radiation therapy. Radiology 158:839–843

Austin-Seymour M, Muzenrider J, Goetin M et al (1986) Proton radiation therapy of chordoma and low grade chondrosarcoma of the base of the skull and cervical spine. Int J Radiat Oncol Biol Phys 12 (Suppl):98

Breteau N, Demasure M, Favre A et al (1996) Fast neutron therapy for inoperable or recurrent sacrococcygeal chordomas. Bull Cancer Radiother 83 (Suppl) 1:142s–145s

Cummings BJ, Hodson DI, Bush RS (1983) Chordoma: the results of megavoltage radiation therapy. Int J Radiat Oncol Biol Phys 9:633–642

Dahlin DC, MacCarty CS (1950) Chordoma: a study of fifty nine cases. Cancer 5:1170–1178

Erlandson RA, Tandler B, Lieberman PH et al (1968) Ultrastructure of human chordoma. Cancer Res 28:2115

Fuller DB, Bloom JG (1988) Radiotherapy for chordoma. Int Radiat Oncol Biol Phys 5: 331–339

Gray SW, Singhabhandhu B, Smith RA, Skandalakis JE (1975) Sacrococcygeal chordoma: report of a case and review of the literature. Surgery 78:573-582

Higinbotham NL, Philips RP, Farr HW, Husti HO (1967) Chordoma: thirty five year study at Memorial Hospital. Cancer 20:1841–1850

Kaiser T, Pritchard D, Onni K (1984) Clinicopathologic study of sacrococcygeal chordoma. Cancer 54:2574–2578

Le Charpentier Y, Bellefqih S, Boisnic S et al (1988) Chordomes. Ann Pathol 8:25–32

O'Neil P, Bell BA, Miller JD, Jacobson J, Guthrie W (1985) Fifty years of experience with chordomas in southeast Scotland. Neurosurgery 16:166–170

Pearlman AW, Friedman M (1970) Radical radiation therapy of chordoma. Am J Roentgenol 108:333–341

Pena CE, Horvat BL, Fischer ER (1970) The ultrastructure of chordoma. Am J Clin Pathol 53:544–551

Raffel C, Wright DC, Guttin PH et al (1985) Cranial chordomas: clinical presentation and results of operative and radiation therapy in twenty-six patients. Neurosurgery 17:703–710

Reddy E, Mansfield C, Hartman G (1981) Chordoma. Int J Radiat Oncol Biol Phys 7:1709–1711

Riche TA, Schiller A, Suit HD, Mankin HJ (1985) Clinical and pathologic review of 48 cases of chordoma. Cancer 56: 82–187

Saunders WM, Castro JR, Chen GTY et al (1981) Early results of ion beam radiation therapy for sacral chordoma. J Neurosurg 64:243

Saxton J (1981) Chordoma. Int J Radiat Oncol Biol Phys 7:913–915

Schmitt G, Schnabel K (1984) Neutron irradiation of soft tissue sarcomas in Essen and Heidelberg. An evaluation of the treatment period 1978-1982. J Eur Radiother 4(3):150–152

Schoenthaler R, Castro JR, Petti PL, Baken-Brown K, Philips TL (1992) Charged particle irradiation of sacral chordomas. Int Radiat Oncol Biol Phys 26:291–298

Spjut HJ, Luce SA (1964) Chordoma: an electron microscopic study. Cancer 17:643–645

Suit HD, Goitein M, Muzenrider J, Verhey L, Davis KR, Koehler A, Linggood R, Ojemann R (1982) Definitive radiation therapy for chordoma and chondrosarcoma of base of skull and cervical spine. J Neurosurg 56:377–385

Tewfik HM, McGinnis WL, Nordstran DG, Latourett HB (1977) Chordoma. Evaluation of clinical behaviour and treatment modalities. Int J Radiat Oncol Biol Phys 2:959–962

Wang CC, James AE Jr (1968) Chordomas: brief review of the literature and report of a case of widespread metastases. Cancer 22:162

Wold LE, Laws ER (1983) Cranial chordomas in childrens and young adults. J Neurosurg 59:1043

Wright D (1967) Nasopharyngeal and cervical chordoma. Some aspects of their development and treatment. J Laryngol Otol 81:1335–1337

Complications of Fast Neutron Therapy

L. Cohen

Neutron Therapy Facility, Fermi National Accelerator Laboratory,
Batavia, IL 60510 USA

Abstract

The purpose of the study was to identify the tissues and organs at risk following high-energy neutron-beam therapy for selected radioresistant tumors, estimating the separate probabilities of both normal tissue injury and of tumor recurrence, each in relation to the absorbed dose.

Published statistical and anecdotal reports on the incidence of serious complications observed following fast neutron treatment directed to the cranium, head and neck, chest, upper abdomen, pelvis, and extremities are reviewed and dose-response parameters derived using bivariate probit or logistic analyses. We then calculate the conditional probability of uncomplicated control (PUC) at various doses, assuming that tumor cure and late injury are stochastically independent events.

The median effective doses and coefficients of variation, derived for neutron irradiation of human brain and spinal cord, oropharynx, lung, stomach and bowel, rectum and bladder, and extremities, are tabulated and tentative "tolerance limits" estimated. Tolerance doses are shown to depend on several factors including beam quality, chemical composition, cell cycling rate, fraction-size, and follow-up time. In patients followed over 5 years, safe tolerance doses appear to range from <14 Gy for the central nervous system up to 22 Gy in the oropharynx and mandible.

Given well-determined dose-response data for specific normal tissues and the associated tumors, the separate probabilities of tumor control and of normal tissue injury at a given dose can be estimated. The particular treatment scheme yielding the highest PUC can usually be identified. The maximum PUC for neutron therapy, compared with other modalities, is a measure of both efficacy and safety for the procedure under study and thus provides a useful guide for comparing various modalities and treatment plans and for designing more effective treatment strategies.

Recent Results in Cancer Research, Vol. 150

Introduction

A central tenet of radiation oncology is that ionizing radiation can ablate malignant tumors with doses that are reasonably well tolerated by contiguous normal tissues. Modern treatment planning and beam delivery systems for photon therapy ensure that any desired dose of low linear energy transfer (LET) radiation can be given to a predefined target volume, which can be shaped to conform to the tumor while excluding uninvolved normal tissues and organs. Tumors that can be treated successfully under these conditions, using safely tolerated doses, have been classified as "radioresponsive."

In the other hand, some human tumors are "radioresistant" and require larger doses, often well beyond normal tissue tolerances. Radioresistance is conferred by various biological adaptations of tumor cells, including intrinsic genetic factors, production of free-radical scavengers, slow cycling with potentially lethal damage repair, tissue hypoxia, or an increased capacity to repair sublethal lessions. Under these circumstances, high-LET radiations (neutrons and heavy ions) may be more effective. Neutrons are primarily indicated for a subset of tumors in which they apparently yield better local control, mainly locally advanced, unresectable, but nonmetastatic radioresistant tumors such as *sarcoma* (bone and soft-tissue), *melanoma,* and *adenocarcinoma* (prostate, salivary gland, thyroid and pancreas).

The newer modality is not without risk. Many older neutron treatment facilities lacked the physical requirements for safe and effective beam delivery. The two modalities can be compared only if dose distributions are comparable, using high-energy (>50 MeV), well-collimated, isocentric neutron beams with skin-sparing, attenuation characteristics, beam-shaping, and treatment plans similar to those obtaining with megavoltage photons. Only then can we determine whether the new modality might yield a higher cure-rate (local control) with the same risk of complications, or fewer complications for the same cure-rate, or both (better control *and* fewer complications).

Reactions and Complications

It is widely believed, probably erroneously, that better control with neutrons can be achieved only at the cost of more late complications. Several factors may have contributed to this impression. Historically, most of the older experience derives from relatively low-energy, cyclotron-generated, poorly collimated, non-isocentric and inadequately blocked or shaped neutron beams. Target volumes were generally larger and less well defined, dosage gradients were less uniform and sensitive normal tissues less well protected, than with modern high-energy photon linacs. Data from these older studies can now be largely discounted. With current facilities risks are minimized but some specific tissues or organs may still be affected depending mainly on the tumor location. Some sites commonly involved by neutron-responsive tumors

Table 1. Tissues at risk in various sites

Tumor type	Location[a]	Tissues at risk
Glioma	Cranium	Brain, eye, ear, hypophysis, scalp, skull
Non-epidermoid carcinoma	Head and neck	Cervical spinal cord, larynx, mandible, salivary glands, temporomandibular joint
Lung cancer	Thorax	Lung, dorsal spinal cord, heart, esophagus, brachial plexus
Pancreatic cancer	Upper abdomen	Stomach, liver, kidney, gut, spinal cord
Adenocarcinoma of prostate	Pelvis	Bladder, rectum, small bowel, hipjoint, sacral plexus
Sarcoma (bone and softtissue)	Extremities	Skin, peripheral nerve, artery, joint

[a] Dose-response functions and tolerance limits are derived for each of these sites.

are listed in Table 1 and are included in reports on neutron injury to brain and spinal cord, mouth and throat, mandible, lung, stomach, bowel and rectum, bladder, peripheral nerve and joints.

Factors Affecting Reactions in Neutron-Irradiated Tissues

Normal tissue reactions to low- and high-LET radiations are qualitatively similar but differ markedly in the doses required to elicit them. The ratio of doses for the same end-point is defined as the relative biological effectiveness (RBE) of the high-LET beam. RBEs are different for different beams (energy spectra), different tissue and tumor types, and various dose or response levels and follow-up times. Biological and dosimetric factors affecting the clinical outcome have been reviewed by Tilikidis and Brahme (1994). The more important physical, biological and technical variables concerned with neutron related complications are listed below.

Energy Spectra of Incident and Scattered Beams

Low-energy beams (or low-energy components in a fast neutron beam) generally have a higher LET (greater RBE) than higher energy particles. Reactions are less severe with "harder" beams (exciting energies around 66 MeV) than with lower energies; similarly neutron beams generated by protons are harder than with deuterons, linac beams are harder, flatter and more uniform (flattening filters not required) than those made in cyclotrons, and thin beryllium targets are better than thick (Petti and Lennox 1994). Also, filters and collimators must be designed to minimize low-energy in-field scatter. The elaborate shielding and collimation used may well abrogate the advantages of "conformal" planning for minimizing normal tissue injury. Due account must also be taken of softening of the beam by internal scatter with increasing depth in tissue.

Biological Models and Dose-Response Relationships

The risk of normal tissue injury by radiation is directly related to dose, time and volume factors. The dose-response relationship is well described by a sigmoid curve with a defined position (radiosensitivity) and slope (variance). Theoretically, a complication is expected to ensue when a functional target population of some billion or so cells is depleted by irradiation to a critically small residual number insufficient to maintain normal structure and function. A stochastic (multitarget) lethality process reduces the surviving clonogenic fraction *(S)* exponentially with increasing dose, eventually reaching a critically small size. Then the probability *(P)* of an irreversible injury (the actual number of surviving cells being less than that required to replenish the depleted population) is given by the Poisson function shown in Fig. 1 (Porter 1980). This sigmoid function is steeper than, but essentially similar in shape to, the macroscopic dose-response relationships observed clinically. Thus the slope and shape of cell survival curves in irradiated tissues depend on both beam quality and tissue-specific cell survival parameters. Genetic determinants of radiosensitivity have also been identified (Busch 1994). Measured RBEs will then differ with different species and with different tissues in the same species. It seems unlikely that useful clinical RBE values could be estimated from experimental animal studies alone; clearly, clinical RBEs must be derived from clinical experience with appropriately fractionated courses of treatment.

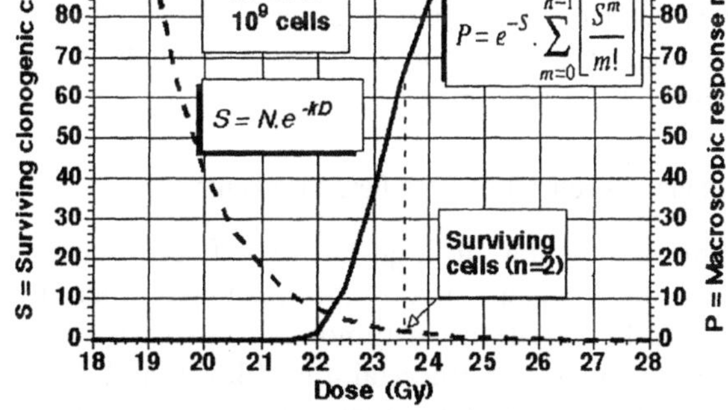

Fig. 1. An exponential or multitarget cell lethality process depletes the initial clonogen population ($N=10^9$ cells) to a critically small surviving fraction *(S)*. The expected complication rate, or the probability of an irreversible injury *(P)*, is then equal to the probability that the actual number of surviving cells is less than the minimum required for repopulation (say, $n=2$ cells). This probability may be calculated by a Poisson exponential summation generating a sigmoidal function of dose

Chemical Composition

Tissue composition and chemical environment both affect RBE estimates. Tissue hydrogen content affects neutron absorption; thus fat, CNS and peripheral nerve may appear relatively sensitive (exhibit *higher* than average RBE values). Dissolved molecular oxygen also affects sensitivity; hypoxia may render some tissues (and tumors) relatively resistant to low-LET radiations and hence also give rise to *higher* estimated neutron RBEs.

Cell Cycle Distribution

With low-LET radiations, cells in the resting or synthetic phases of the mitotic cycle appear less sensitive than mitotic or premitotic cells. With low LET, slowly cycling tissues may permit considerable repair of potentially lethal damage during this insensitive phase. The phase-related difference is much less marked with high-LET radiation. Consequently, the observed neutron RBE will be *greater* in slowly cycling tissues and tumors having a larger proportion of noncycling or resting cells.

Dose Level, Fraction Size, End-Point, and Level of Effect

Low-LET radiation generally yields a convex cell survival curve with a relatively small initial slope and a steeper terminal portion. Therefore most tissues are relatively insensitive at small doses or small fractions. With high-LET radiations, initial (low-dose) slopes are always steeper while the terminal (high-dose) slope is less affected. A direct consequence of the *difference* in curvature of cell survival functions for photons and neutrons is that RBE is greatest at low doses and diminishes with increasing dose (Gueulette et al. 1984; Wambersie et al. 1982). The RBE for a single exposure ranges between two asymptotic levels, corresponding to the ratio of initial slopes (lowest dose, highest RBE) and the ratio of terminal slopes (highest dose, lowest RBE), respectively (Fig. 2). For fractionated radiation therapy the effect is determined by individual fraction sizes. In clinical practice neutron RBEs are defined by comparison with "conventionally fractionated" photons (generally under 2 Gy/fraction).

Shape (Steepness) of Dose-Response Function

The ratio of low- to high-LET iso-effective doses for two modalities depends also on the shape of the dose-response function (Fig. 3). This in turn is affected by variations in tumor-cell sensitivity as well as dosimetric heterogeneity, errors and uncertainties (Boyer and Schultheiss 1988). The dose-response curve is often measurably steeper for neutron-irradiated normal tissues, in which case the estimated iso-effective RBE varies with the response level studied.

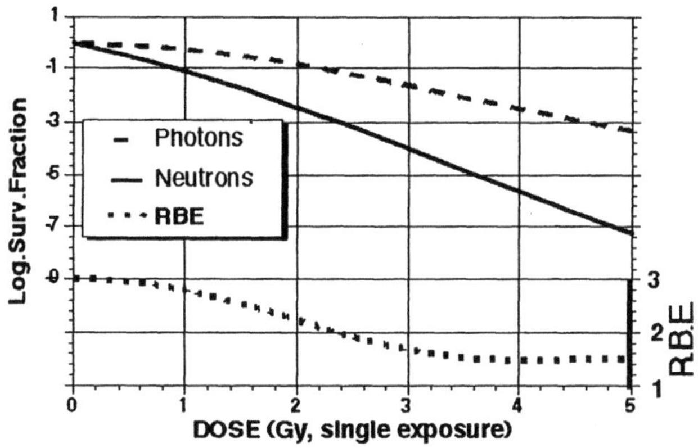

Fig. 2. Relative biological effectiveness *(lowest line)* of a neutron beam compared to photons for normal tissue injury. Cell-survival curves are shown for photons *(upper broken line)* and high-energy neutrons *(solid line)*. The figure is redrawn from Wambersie et al (1982) using a cell kinetic model and parameters taken from prior published data. The estimated RBE ranges from about 3.0 at small fractional photon doses to an asymptotic minimum of 1.5 for large doses

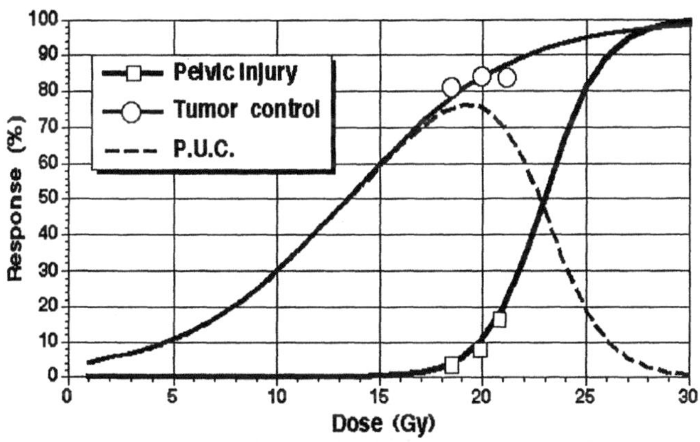

Fig. 3. A typical six-point assay comparing local control of late-stage carcinoma of prostate *(left-hand curve)* and grade 3 injury to contiguous pelvic organs (bladder and rectum, *right-hand curve)*, using 66 MeV (Fermilab) neutrons (Cohen et al. 1995) at three dose levels: 19, 20.4, and 21 Gy. Tumor control (about 83%) varied little with dose; normal tissue complications increased steeply with increasing dose. The estimated maximum probability of uncomplicated control *(broken line)* appears to be about 76% at 19 Gy

Follow-up Time and Censoring

Late effects are irreversible and cumulative; they will consequently appear to increase as follow-up continues (Fig. 4). If a new modality yields better overall survival, there will be more long-term survivors at risk for developing complications. However, the total number of survivors decreases with time and the absolute prevalence of complications will also diminish with attrition of the surviving population. For a true estimate of late effects, and to pro-

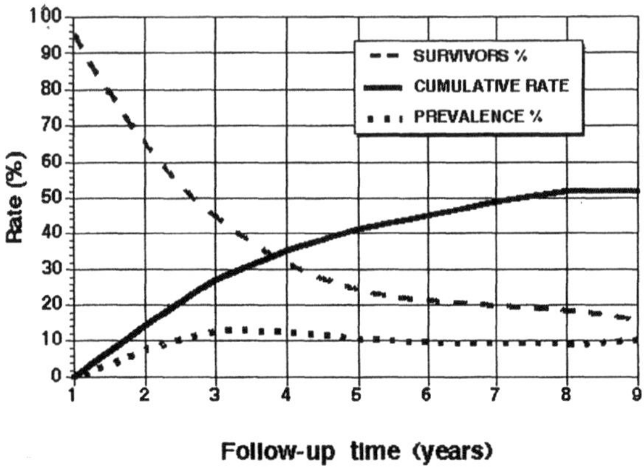

Fig. 4. Complication rates in a series of locally advanced radioresistant tumors treated at Fermilab and followed for 9 years (Cohen et al. 1989). Long-term survival *(broken line)* is under 20%; at any given time the observed prevalence of complications *(dotted line)* is around 10% of the survivors; the calculated actuarial cumulative incidence of late injuries (adjusted for competing mortality) reaches 50% by 9 years. Clearly, observed complication rates will depend on the follow-up time and the longevity of the population studied

vide a fair comparison with photons, estimated complication rates must be corrected for censoring due to intercurrent morbidity and mortality (Schultheiss et al. 1986).

Observed Reactions and Injuries in Specific Organs

The clinical literature contains a number of reports on reactions following various doses of neutrons delivered to each of the six anatomical regions described in Table 1. These are sufficient to estimate (at least to a first approximation) the median doses and variances for complications at the relevant sites (Table 2). In addition, there are many sporadic reports which are insufficient for detailed analysis but serve to identify specific tissues and organs at risk. Some of these observations are described below; it should be noted that these reactions are uncommon and largely avoidable.

Cranium

Skin changes and hair loss are commonly observed in the irradiated scalp but are of little long-term import compared with the much greater risks associated with intracranial structures. The *brain* is extraordinarily sensitive to neutrons. Early studies on the treatment of brain tumors (with relatively low-energy beams) showed that doses in excess of 16 Gy often led to rapidly fatal encephalopathy (Catterall and Bewley 1979; Griffin et al. 1983). Lower doses were better tolerated, though risks were hard to assess because of competing mortality from advancing tumor and the difficulty in evaluating such non-lethal effects as loss of cognitive function. The location and extent of mini-

Table 2. Observed tolerance limits for neutron irradiation. (Cohen et al. 1989; Schultheiss et al. 1990)

Site	Organ Parameters[a] →	Crude		Adjusted		Estimated tolerance
		μ	σ	μ	σ	(μ–σ) Gy
Cranium	Brain and spinal cord	–		–		11–14
Head and neck	Skin, oropharynx, mandible	33.4	11.0	25.7	3.6	22.1–22.4
Thorax	Lung	29.3	8.2	20.9	3.0	17.9–21.1
Abdomen	Stomach and bowel	31.0	8.0	23.7	4.7	19.0–23.0
Pelvis	Rectum and bladder	29.4	9.9	21.1	2.9	18.2–19.5
Other	Extremities	22.0	3.0	–		19.0
Composite	All sites	29.0	7.9	–		21.1

[a] High-energy (66 MeV), 5-year follow up, μ=median doses (ED_{50}), σ=standard deviations (slopes) of probit functions. Doses in gray.

mal neutron-induced encephalopathy may be determined by magnetic resonance imaging (MRI) scans which readily demarcate heavily irradiated segments from unaffected brain. The *temporal lobe* of the cerebral cortex is often involved in treatment of tumors of the upper respiratory tract (Errington 1986; Griffin et al. 1986); other cranial sites at risk are the inner *ear*, the *pituitary* gland and optic *chiasm*, and the *eye* including lacrymal gland, cornea, retina and optic nerve.

Head and Neck

The cervical *spinal cord* is inevitably irradiated during radical treatment of cancer of the upper aerodigestive tract and associated lymph nodes, and meticulous treatment planning is required to minimize the risk of cord injury. With high-energy beams and standard fractionation (12 fractions over 4 weeks), doses not exceeding 13 Gy at the central cord isodose line are believed to be safe (Cohen et al. 1985a). This is an important constraint for treatment planning; set too high this limit could permit the development of a devastating complication such as paraplegia, but set too low it might compromise the attainment of an adequate tumor dose and thus increase the equally devastating risk of recurrent cancer (Peters et al. 1986).

The primary target volume may require doses of about 21 Gy for adequate local control. At this dose level *salivary gland* function is suppressed, leading to disabling xerostomia, dysphagia and dental caries (Maor et al. 1995) unless treatment plans are designed to spare at least some secretory tissue; there is a transient loss of *taste*, occasional fibrosis and edema of *larynx* and *oropharynx* (Griffin et al. 1986), acute and late *skin* reactions (including epilation and telangiectasia); and suppression of *thyroid* function. The flat bones of the skull generally tolerate neutrons well; the *mandible* may be an exception because of the risk of infection by way of tooth sockets or overlying mucosal ulceration (Catterall 1986).

Thorax

Some normal *lung* is always included in the treatment of intrathoracic cancer. A late consequence of this is the appearance on chest radiographs or scans of localized fibrosis conforming in position and shape to the treatment portals (Griffin et al. 1986). This may be asymptomatic if confined to a limited target volume. If a substantial neutron dose reaches the *esophagus*, fairly severe consequences, including acute symptomatic esophagitis, stricture and perforation may result with life-threatening leakage into the bronchial tree, pleural space or mediastinum.

Abdomen

Most neutron injuries in the upper abdomen have followed attempts to irradiate unresectable pancreatic cancer. Planning contours through the pancreas necessarily encompass liver, lumbar spinal cord, kidneys, stomach and intestine. With target absorbed doses of 20 Gy or more, late effects in the liver (usually asymptomatic) may appear and can be detected by liver scans showing localized areas of diminished uptake and, less frequently, more severe consquences due to portal veno-occlusive disease (Cohen et al. 1985b). Neutron-related lumbar myelopathy and nephropathy are theoretical possibilities but, since both spine and kidneys are easily excluded from the target volume, have not been observed in practice. The most significant complications are observed in the *stomach;* post-irradiation acute gastritis, late gastric ulceration, bleeding and perforation are limiting side effects of treatment to this area (Cohen et al. 1996).

Pelvis

In treating prostatic cancer, the high-dose target receives some 19–21 Gy and includes the contiguous anterior *rectal wall* and posterior *bladder.* The risk of injury is both dose and volume dependent (Schultheiss et al. 1983, 1995). At higher doses (around 22 Gy prescribed for chondrosarcoma of pelvic bone, for example) radiation injury with stricture and perforation of the *small intestine* has been observed. Skin reactions, followed by atrophy and fat loss with subcutaneous fibrosis localized to the treatment portal, are common sequelae of lower abdominal neutron irradiation (especially in the fat, pendulous abdomen). *Lymphedema* of scrotum and lower extremities, sensory and motor loss from injury to *sacral plexus,* and stiffness due to periarticular fibrosis of *hip joints* are some of the less common sequelae.

Extremities

Neutrons may be used for treatment of melanoma or sarcoma of the extremities delivering relatively high doses (around 22 Gy). In contrast to photons, neutron absorption in *bone* is less than in soft tissue so even at this large dose osteitis and bone necrosis are not common problems. However, *fibrosis* of soft tissues (fascia, muscle, fat, synovium) may be severe. Though seldom life-threatening, lymphedema of the limb, restriction of motion by periarticular fibrosis, ischemia due to arterial obliteration, and sensory and motor loss from peripheral neuropathy can combine to provide severely disabling consequences.

Dose-Response Assays

In order to estimate safe and effective doses for a new modality, and to compare these estimates with conventional experience, fairly precise dose-response assays are required. Both tumor-control and normal tissue complication probabilities may be described by sigmoid curves (Fig. 5), fitted to observational data by Poisson, probit or logistic statistical models (Table 3). The *Poisson* function is calculated on the basis of known or assumed cell survival parameters for the target cell population, assuming that depletion below a predetermined residual level leads to irreversible injury (Fig. 1). This approach generates theoretical sigmoid response curves for complications considerably steeper than those observed clinically. Dosimetric gradients, errors and uncertainties, as well as intrinsic variation in cellular radiosensitivity, make for shallower curves. By contrast, the *probit* function is empirically fitted to observational data using three parameters, respectively, for the median position (μ), spread (σ) and confidence limits (χ^2) of the relationship (Fig. 6). The *logistic* formula can be similarly fitted, adjusting the median (μ), slope (γ) and goodness-of-fit parameters to the data (Schultheiss et al. 1990).

Table 3. Sigmoid dose response functions

Probit	$P = \frac{1}{\sigma\sqrt{2\pi}} \int_{-\infty}^{D} e^{(D-\mu)^2} \cdot dD$ $\approx \frac{1}{2} \cdot \{1 - erf[\sqrt{\pi} \cdot \gamma \cdot [1-(D/\mu)]]\}$
Logistic	$P = \frac{1}{[1+(D/\mu)^{4\gamma}]}$
Poisson	$S = N \cdot e^{-kD}$ $P = e^{-S} \cdot \sum_{m=0}^{n-1} \left[\frac{S^m}{m!}\right]$

P is the probability of irreversible late tissue damage. μ is the empirical (observed or extrapolated) median for 50% response. σ represents the spread (standard deviation) of the gaussian or probit distribution, and is inversely related to γ the empirical slope (dP/dD) of the steepest portion of the curve. In the Poisson model, S is the surviving population after D Gy is delivered to N cells, k is a cellular radiosensitivity parameter, and n is the critical mass or "tissue-rescuing" cell number.

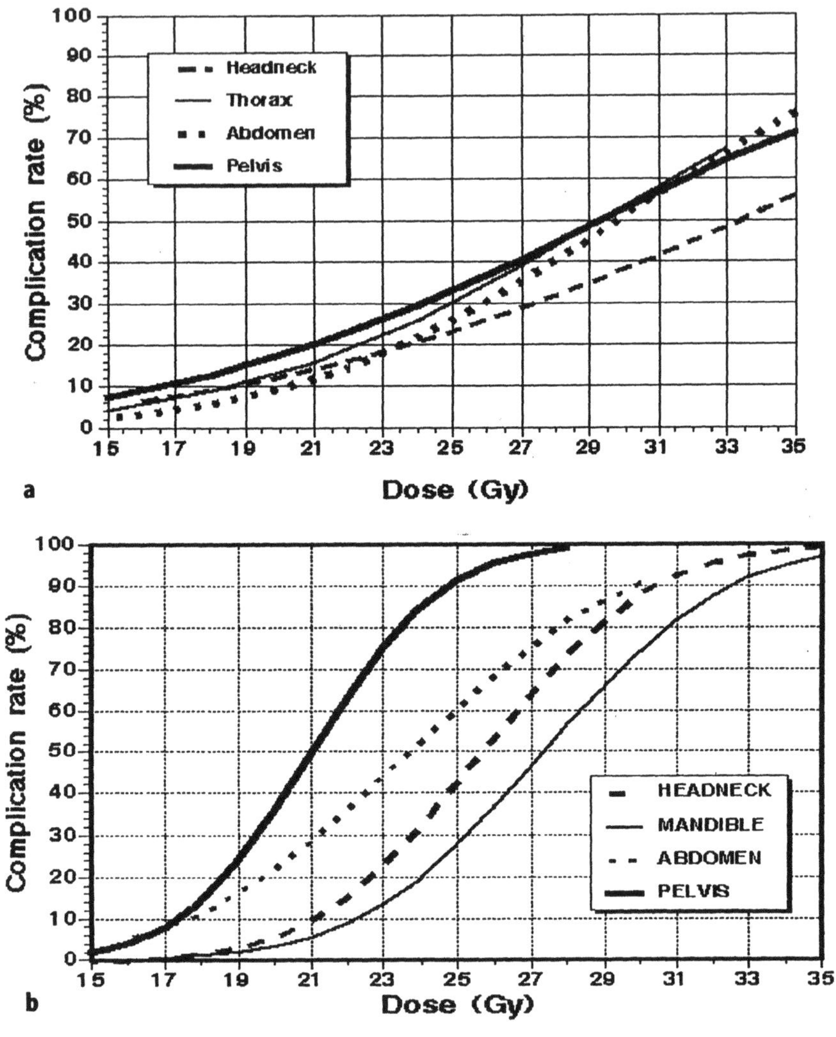

Fig. 5 a, b. Dose-response curves fitted to the observed or adjusted incidence of complications for anatomical regions in which two or more data points were available for analysis. **a** Crude rates, directly observed and counted (Cohen et al. 1989), give relatively flat response curves with large errors and uncertainties; medians, estimated by extrapolation, appear to cluster around 29 Gy, and coefficients of variation are large (σ/μ about 25%). **b** When adjusted for censoring (Schultheiss et al. 1990) corresponding data yield steeper, more clearly sigmoid, dose-response curves, estimated medians are lower 21–27 Gy) and coefficients of variation are small (<15%)

Clinical observations of complication rates are usually based at the lower end of the scale (doses well below the median value) so that estimated medians are always extrapolations beyond the range of experience. A better-defined and more convenient parameter is the 16% risk level (μ–σ, where uncertainty limits are narrower) rather than the median (μ). Some estimates of

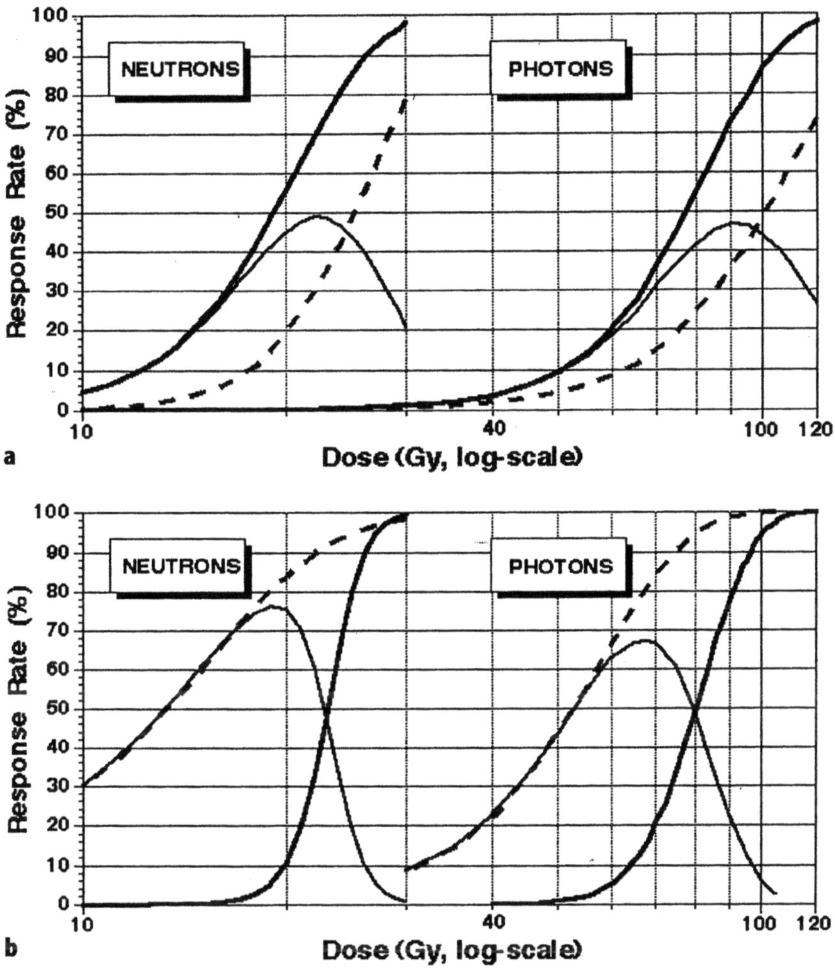

Fig. 6a, b. Examples of four-curve assays (8–12 data points) for optimization of a new modality. **a** Early data (Cohen 1982; Maor et al. 1995) for neutron therapy in locally advanced epidermoid carcinomas of head and neck together with photon controls, fitted by probit analysis to sigmoid dose-response functions. The four curves are almost parallel (using a logarithmic dosage scale). The RBE of the beam is similar for both tumor and normal tissue and the maximum probability of uncomplicated control (PUC) is about the same (<50%) for both modalities. **b** A more recent nonrandomized study on locally advanced (T3–T4) prostatic cancer. The neutron data are those reported by Fermilab (Cohen et al. 1995); the photon data are taken from contemporary literature. The four curves are probably not parallel. The RBE is about 3 for the normal pelvis and nearer 4 for the tumor; the maximum PUC is higher (76%) for neutrons than for photons (66%).

these parameters for high-energy neutrons are given in Table 2. With either the probit or logistic empirical models, two data points are sufficient to estimate the position (radiosensitivity) and slope (variance) of a given curve. Three or more points are needed to provide a measure of the robustness or confidence limits of these estimates. A minimum of six points (three for each procedure) would be required to compare two different modalities, and a four-curve assay (12 data points defining tumor and normal tissue responses for both photons and neutrons) is required to display complete therapeutic paradigms (Fig. 6).

A major advantage of this type of assay is that optimization becomes feasible along the lines originally suggested by Holthusen (1936) and Strandqvist (1944). Ideally, this process requires that an optimal target absorbed dose as well as an optimal treatment plan be used. Doses can then be selected so as to maximize the conditional probability of tumor control without complications (Fig. 3). Since the tumor dose-response curve tends to be less steep than that for normal tissue injury, the computed probability of uncomplicated control (PUC) is likely to be somewhat skewed (Cohen et al. 1995). This implies that even a small overdose could yield an unacceptably high risk of injury, while correspondingly small underdosage could be almost as effective but considerably safer. Given a near-optimal neutron dose, optimizing the treatment plan (using conformal target volumes, for example) can reduce complication risks to the level observed with best current practice in conventional radiotherapy (Schultheiss et al. 1995). The four dose-response curves (8–12 data points) in Fig. 6 allow neutron and photons to be compared when both are given at optimal dose levels.

References

Boyer AL, Schultheiss T (1988) Effects of dosimetric and clinical uncertainty on complication-free local tumor control. Radiother Oncol 11:65–71

Busch D (1994) Genetic susceptibility to radiation and chemotherapy injury: diagnosis and management. Int J Radiat Oncol Biol Phys 30:997–1002

Catterall M (1986) The late effects on normal and malignant tissues from 7.5 MeV neutrons in the treatment of advanced mouth cancer. Bull Cancer (Paris) 73:587–590

Catterall M, Bewley DK (1979) Fast neutrons in the treatment of cancer. Academic, London; Grune and Stratton, New York

Cohen L (1982) Absence of a demonstrable gain factor for neutron beam therapy of epidermoid carcinoma of the head and neck. Int J Radiat Oncol Biol Phys 8:2173–2176

Cohen L (1986) Calculation of clinical r.b.e. values or neutrons. Int J Radiat Biol 50:147–154

Cohen L, Ten Haken RK, Mansell J, Yalavarthi S, Hendrickson FR, Awschalom M (1985a) Tolerance of human spinal cord to high energy p(66)Be(49) neutrons. Int J Radiat Oncol Biol Phys 11:743–749

Cohen L, Woodruff K, Hendrickson F, Kurup PD, Mansell J, Awschalom M, Rosenberg I, ten Haken RK (1985b) Response of pancreatic cancer to local irradiation with high energy neutrons. Cancer 56:1235–1241

Cohen L, Schultheiss TE, Hendrickson FR, Mansell J, Saroja KR, Lennox A (1989) Normal tissue reactions and complications following high-energy neutron beam therapy. I. Crude response rates. Int J Radiat Oncol Biol Phys 16:73–78

Cohen L, Saroja KR, Hendrickson FR, Lennox AJ, Hatcher MA, Kroc TK (1995) Neutron irradiation of human pelvic tissues yields a steep dose-response function for late sequelae. Int J Radiat Oncol Biol Phys 32:367–372

Cohen L, Hendrickson FR, Lennox AJ, Kroc TK, Hatcher MA, Bennett B (1996) Pancreatic cancer: treatment with neutron irradiation alone and with chemotherapy. Radiology 200:627–630

Errington RD (1986) Advanced carcinoma of the paranasal sinuses treated with 7.5 MeV fast neutrons. Bull Cancer (Paris) 73:569–576

Griffin TW, Davis R, Laramore G, Hendrickson F, Rodrigues-Antunez A, Hussey D, Nelson J (1983) Fast neutron radiation therapy for glioblastoma multiforme: results of an RTOG study. Am J Clin Oncol 6:661-667

Griffin TW, Pajak T, Laramore G, Davis L (1986) Analysis of neutron radiotherapy treatment complications. Bull Cancer (Paris) 73:582–586

Gueulette J, Octave-Prignot M, Wambersie A (1984) RBE of d(50)+Be neutrons, as a function of the dose per fraction, for early intestinal tolerance in mice. J Eur Radiother 5:180–182

Holthusen H (1936) Erfahrungen über die Verträglichkeitsgrenze für Röntgenstrahlen und deren Nutzanwendung zur Verhütung von Schäden. Strahlentherapie 57:254–269

Maor MH, Errington RD, Caplan RJ, Griffin TW, Laramore GE, Parker RG, Burnison M, Stetz J, Zink S, Davis LW, Peters LJ (1995) Fast-neutron therapy in advanced head and neck cancer: a collaborative international randomized trial. Int J Radiat Oncol Biol Phys 32:599–604

Peters LJ, Schultheiss T, Maor MH (1986) Tolerance of human spinal cord to high energy neutrons (Letter to the Editor). Int J Radiat Oncol Biol Phys 12:292–293

Petti PL, Lennox AJ (1994) Hadronic radiotherapy. Annu Rev Nucl Part Sci 44:155–197

Porter EH (1980) The statistics of dose-cure relationships for irradiated tumours. Br J Radiol 53: 210–227, 336–345

Schultheiss TE, Orton CG, Peck RA (1983) Models in radiotherapy: volume effects. Med Phys 10:410–415

Schultheiss TE, Thames HD, Peters LJ, Dixon DO (1986) Effect of latency on calculated complication rates. Int J Radiat Oncol Biol Phys 12:1861–1865

Schultheiss TE, Cohen L, Mansell J (1990) Normal tissue reactions and complications following high energy neutron beam therapy. II. Complication rates adjusted for censoring. Int J Radiat Oncol Biol Phys 18:165–171

Schultheiss TE, Hanks GE, Hunt MA, Lee WR (1995) Incidence of and factors related to late complications in conformal and conventional radiation treatment of cancer of the prostate. Int J Radiat Oncol Biol Phys 32: 643–649

Strandqvist M (1944) Studien über die kumulative Wirkung der Röntgenstrahlen bei Fraktionierung. Acta Radiol Suppl 55

Tilikidis A, Brahme A (1994) Microdosimetric description of beam quality and biological effectiveness in radiation therpay. Acta Oncol 33:457–469

Wambersie A, Gueulette J, Dutreix J (1982) Shape of the initial part of the cell survival curve. Implications for the RBE/Dose relationship for small doses per fraction. In: Kärcher KH, Kogelnik HD, Reinartz G (eds) Progress in radio-oncology II. Raven, New York pp 99–104

Is There a Role for Heavy Ion Beam Therapy?

J. Debus[1,2], O. Jäckel[1], G. Kraft[3], and M. Wannenmacher[2]

[1] Department of Radiation Oncology, German Cancer Research Center (DKFZ), Im Neuenheimer Feld 280, D-69120 Heidelberg, Germany
[2] Department of Radiotherapy, University Hospital, Rupprecht-Karls-University, Im Neuenheimer Feld 300, D-69120 Heidelberg, Germany
[3] Gesellschaft für Schwerionenforschung (GSI), Planckstrasse 1, D-64291 Darmstadt, Germany

Abstract

The aim of this contribution is to review the radio-oncological rationale of heavy ion beam radiotherapy in the management of cancer. Protons and helium ions are being investigated because of the improved dose distributions, perhaps superior in many clinical situations to those obtainable with photons or electrons. Heavy ions also bear the advantage of superior dose distribution and may additionally provide higher biological effectivity. A substantial database of historical results supports the hypothesis that conformal radiotherapy is superior to conventional radiotherapy. For the selection of patients for ion beam therapy, the following questions have to be evaluated: Can any significant radiation morbidity by conventional treatment expected? Is the radiation morbidity caused by unintended irradiation of non-target tissue outside the planning target volume? Can the tumor control be improved due to dose escalation with less radiation morbidity? A substantial number of patients have been treated by light ion radiotherapy. There are only a few clinical trials attempting to compare conventional photon radiotherapy with light ion radiotherapy. Clinical results with heavy ions such as carbon ions seem very promising in certain tumors. However results of randomized trials are still to be seen. Clearly the control arm has to be designed to be as close as possible to the optimal photon/electron treatment methods in use. The feasibility of heavy ion treatment has been demonstrated worldwide, with more than 10000 patients. In the near future more centers worldwide will start patient treatments. The wider availability of light ion beams for clinical use will enable the establishment of a database of clinical results and the elucidation of the role of heavy ion beams in the treatment of cancer.

It has been shown in previous chapters that there may be a clinical benefit of high-linear energy transfer (LET) radiotherapy compared to low-LET radiotherapy. The aim of this contribution is to review the radio-oncologic rationale of ion beam radiotherapy in cancer management.

Recent Results in Cancer Research, Vol. 150

The Radio-oncological Rationale for the Study of the Application of Ion Beams in the Management of Cancer

Cure rates for patients with various malignancies have improved over recent decades due to advances in the field of radiation oncology. The clinical results for surgery alone, photon radiation therapy or combined modality treatments are good and improving for cancers of many sites. Nevertheless there are still about 30% of all cancer patients with localized disease where failure in the treatment of the primary tumor contributes materially to death. It has been predicted that an improvement in local control of 50% will improve survival rates by about 10%–15% (Suit 1988). Intensified radiotherapy may be one potential tool to improve local control. The superior dose deposition of ion beams compared to photon beams was recognized a long time ago. Since 1957 more than 10000 patients have been treated with ion beam radiotherapy (Suit 1992). Protons and helium ions are being investigated because of their improved dose distributions, which might be superior in many clinical situations to those obtainable with photons or electrons. Heavy ions also have a dose distribution advantage and may additionally show a biological advantage due to their higher biological effectivity.

Conforming the Treatment Volume to the Target Volume

The physical basis for the interest in the application of ion beams to radiation oncology is the potential for improving radiation dose distributions (Goitein 1992). The superior dose distributions of ions are a result of the limited range of charged particle beams in matter due to the stopping process. In contrast the intensity of photons is reduced exponentially. The aim of conformal radiotherapy is to minimize the difference between the treatment volume and the planning target volume. The tolerance to radiation of most of the organs at risk depends on the volume of normal tissue which is treated. As a consequence of conformal techniques, normal tissue complication probability (NTCP) is reduced and tumor control may be improved depending on the total dose prescribed and the fractionation.

There are two categories of treatment-related morbidity due to radiation therapy: the acute and subacute radiation reactions and so-called late delayed radiation reactions. While acute radiation reactions occur between days and a few months after radiotherapy, late delay radiation reactions may occur some years after radiotherapy. The natural history of acute subacute radiation reactions is often self-limiting, while late delayed radiation reactions tend to be progressive. These late delayed reactions are often considered as complications which are generally unintended. It is a frequent clinical finding that radiation treatment related morbidity may develop in nontarget tissues. The improvement of the dose distribution allows the reduction of the frequency and severity of treatment-related morbidity.

A variety of approaches are used to conform the dose to the target volume by different techniques. At a very early stage brachytherapy was optimized by placing the radiation sources in the tumor. Photon therapy has experienced tremendous progress due to three-dimensional treatment planning and techniques such as inverse treatment planning and intensity-modulated radiotherapy. Ion beams are another method of delivering conformal radiotherapy.

The reduction of the treated volume is essentially a physical problem, which should have no influence on the tumor response as long the planning target volume is defined correctly. Therefore conformal therapy requires very exact definition of the planning target volume based on an oncological knowledge of the spread of the tumor and referral to modern imaging techniques. The treated volume, which is defined as the volume receiving a significant portion of the dose, can be reduced by conformal techniques, and organs at risk can be spared. It has been shown for several critical structures that reducing the volume included in the high-dose area results in better radiotolerance (Debus et al. 1997). However, it has be kept in mind that organs at risk that are included in the planning target volume may not be spared because they are considered to be contaminated by tumor cells. Sparing of organs at risk within the target volume may therefore jeopardize tumor control. This situation can be found in the treatment of the carcinoma of the prostate, where the anterior rectal wall is part of the target volume.

Improving Radiobiological Effectivity Due to the High-LET Effect

The radiobiological effectivity of proton beam therapy is nearly equal to that of conventional high-voltage photon radiotherapy. Therefore from the radiobiological point of view there is not a substantial difference between proton beam and photon beam therapy. The clinical relative biological effectiveness (RBE) is considered to be 1.1. But as the beam quality changes to high-LET radiation the biological effectivity of the beam increases. Biologically fast neutrons and heavy ions such as carbon and neon increase the RBE.

It is known from early fast neutron radiotherapy that high-LET radiotherapy may improve local control in selected indications, which are shown in Table 1. Overall the number of prospectively randomized trials comparing high-LET with low-LET radiotherapy is small.

High-LET radiation has some biological advantages compared to conventional low-LET radiation as discussed in this book. The radiobiological effects include:

- Less influence of the oxygen concentration in the tissue
- Less repair/high α- to β-ratio
- Less variation of radiosensitivity through the cell cycle

1. Historically the rationale for introducing fast neutrons in therapy was the existence of hypoxic cells and a reduction of the oxygen enhancement ratio

(from about 3 to 1.8) (Field and Fowler 1984). It has been shown that in clinical situations hypoxic cells could be responsible for local tumor radioresistance. These data provide support for techniques specifically directed towards the killing of hypoxic cells, which of the high-LET radiations are probably the most promising techniques together with local tumor hyperthermia (Griffin et al. 1994).

2. Cells from tumors as well as normal tissue show a large variation in the sensitivity to ionizing irradiation. These differences are enhanced by the capacity for repair of sublethal damage. Furthermore for a given cell line, the radiosensitivity varies with the position in the, mitotic cell cycle. High-LET radiation produces lethal damage to the DNA with greater effectiveness and accumulation of sublethal damage contributes less to the total biological effect. Repair phenomena then play a less important role (Hall and Kraljeric 1976). The large variation observed between RBE values of different cell lines is due to a greater extent to differences in the responses to photons rather than to high-LET radiation.

In summary all cell types in all conditions show less variation in radiosensitivity to high-LET radiation than to low-LET radiation.

Which Patients Are Considered to Profit from Precise Dose Localization?

From the physical point of view conformal radiotherapy is possible in all patients treated and therefore the side effects may be reduced. Whether the patient benefits from conformal treatment depends on the clinical situation. The benefit is potentially higher the more treatment-related morbidity outside the target volume is caused by conventional techniques.

The technical feasibility of delivering precise dose distributions to the patient depends on an exact definition of the whole target volume. It is a crucial step for the whole therapy to outline the target volume based on three-dimensional image information including a clinical knowledge of the tumor spread. Therefore modern three-dimensional treatment planning is a prerequisite for successful conformal treatment (Goitein 1992). If the target is chosen too large, the risk of radiation morbidity may rise and thereby limit the total deliverable dose. Another concern with the utilization of the smaller treatment volumes which may be achieved by ion beam therapy is the potential for a higher rate of failure in eradicating the tumor because some portion of the margin of the tumor was not included in the target volume. Of relevance here is the fact that during recent decades treatment volumes have been reduced with the introduction of high-energy units, simulators, CT and MRI-based treatment planning, etc. These progressively smaller treatment volumes have been accompanied by fewer marginal misses. But it has to be considered that in the past the treatment volume was largely based on the definition of the structures which were unambiguously nontarget. This was derived from the knowledge of patterns of local spread and the sites of failure of both radiation therapy and surgery. However, exclusion of nontarget tissues/structures from the treatment volume does not result in an increased rate of marginal failure. Application of ion beam therapy techniques to certain tumor categories will require lower treatment volumes. The clinical gain will be an increase in the proportion of patients with local cure without complication. There is also the potential with certain lesions to reduce the scope of surgery or to replace surgery by the use of more effective radiation therapy. For example, it has been shown that stereotactic single high-dose irradiation ("radiosurgery") of brain metastases achieves similar tumor control to neurosurgical intervention (Engenhart et al. 1993). It was demonstrated that the treatment-related morbidity of radiosurgery was considerably lower than neurosurgical excision. About 90% of the brain metastases could be locally controlled after radiosurgery, whereas with conventional whole brain irradiation, only 45% of brain metastases were controlled. This example shows how tumor control can be increased by dose escalation after conforming the treatment volume to the target volume. The problem of radiosurgery alone is that microscopic disease that is not detectable by imaging techniques remains untreated. It is known that about 20% of patients treated with radiosurgery alone will develop further brain metastases during the follow-up (Engenhart et al. 1993). Currently it is under evaluation whether these patients will benefit from additional whole brain irradiation.

The dependence of patient tolerance of radiation on the volume of normal tissue and the proportion of a critical normal structure in the treatment volume is less well studied than the effect of dose. There are only few quantitative data available (Debus et al. 1997). Different organs have different tolerances regarding the volume of tissue irradiated. However, regardless of dose/volume effects there is no advantage to the patient in the treatment involving irradiation of clearly uninvolved tissue when there are technical means of excluding this tissue from the treatment volume.

The following evaluations are required for the selection of patients for ion beam therapy:

1. Whether any significant radiation morbidity by conventional treatment can be expected.
2. Whether the radiation morbidity is caused by unintended irradiation of nontarget tissue outside the planning target volume?
3. Whether the tumor control can be improved due to dose escalation with lower radiation morbidity.

Moreover, patients with known increased susceptibility to radiation-induced side effects are considered to profit from conformal treatment. This reduced radiotolerance could be due to various reasons. One of the most common reasons is previous radiotherapy. Therefore patients with locally recurrent tumors after definitive radiotherapy may have an additional therapeutic option due to conformal treatment.

Patients with constitutionally caused reduced radiotolerance such as little children requiring treatment of brain tumors may profit from this kind of treatment (Gademann and Wannenmacher 1992). Moreover, modern molecular biology may one day be able to which predict patients have genetically determined reduced radiotolerance as is found in ataxia telangiectasia.

The most common use of ion beam radiotherapy is for tumors located directly next to an organ at risk which might develop an intolerable radiation reaction, for example, a chordoma of the clivus, where the brain stem, the optic chiasm and the pituitary gland surround the target volume.

It has been mentioned earlier that inverse treatment planning and intensity-modulated radiotherapy are able to generate highly conformal treatment plans. A typical example is shown in Fig. 1, which compares the three-di-

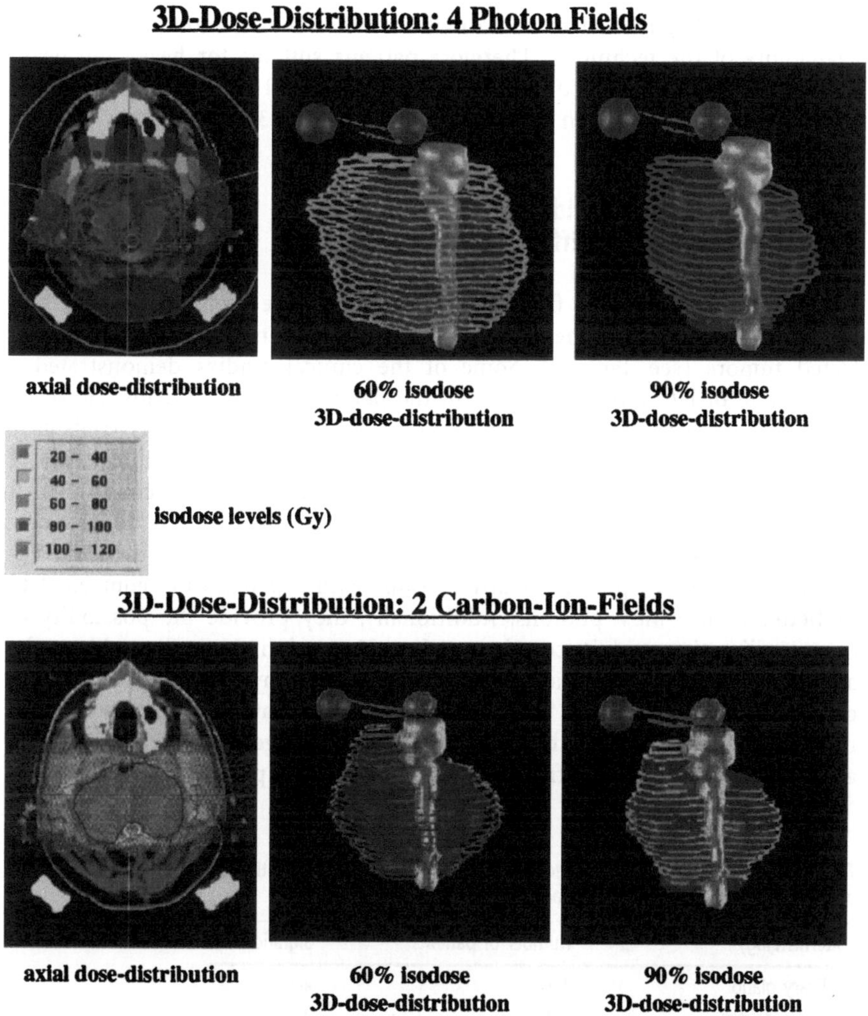

Fig. 1. Patient with a base of skull chordoma. Comparison of the three-dimensional dose distribution of an optimized four-field photon technique and a two-field carbon ion treatment with volume scanning. It can be easily recognized that the ion treatment results in better conformation of the dose to the target

mensional dose distribution of the four-field photon technique with a two-field carbon ion technique. Therefore it needs to be evaluated on a scientific basis whether modern photon treatment techniques can achieve the required dose distribution using sophisticated treatment planning with lower costs than ion beam therapy.

Results

Prospectively randomized trials have shown that ion beam therapy of chordomas and chondrosarcomas of spine and skull is superior to conventional radiotherapy. Application of conformal radiotherapy depends greatly on the availability of the technique. Therefore patients suitable for heavy ion beam therapy should be stratified according to the expected benefit from ion beam therapy. A system has been proposed by Drumm (1993).

Which Patients Are Considered to Profit from Higher Biological Effectiveness?

The survival and freedom from treatment failure have been significantly better with high-LET radiotherapy than with conventional radiotherapy in selected tumors (see Table 1). Some of the clinical studies demonstrated a higher rate of late radiation injury in the neutron treatment arms. However, it has to be considered that the neutron generators used in earlier trials gave considerable problems concerning performance. Many of the clinical data at present available were obtained under "suboptimal" technical conditions. It was shown by Griffin et al. (1993) that the rate of side effects is directly correlated to the physical selectivity of the neutron generator. High-energy cyclotrons (>40 MeV) provide neutron beams with a penetration equivalent to or better than 6-MeV photons. Additionally, they provide the possibility of beam collimation and isocentric rotation. Heavy ion therapy combines the advantage of high-LET radiotherapy with the ideal physical characteristics of conformal dose deposition. Therefore patients who have tumors which might respond to high-LET radiotherapy and where this target volume is next to organs at risk might benefit from heavy ion beam therapy.

Table 1. Clinical trials which revealed and advantage of high-LET radiotherapy over conventional photon/electron radiotherapy (Griffin et al. 1994)

Site/histology	Number of patients	Significance
Salivary gland	150	++++
Prostate cancer	285	+++
Bone sarcomas	105	++
Soft tissue sarcomas	540	++

The number of patients suitable for each radiotherapeutic approach has been described by Drumm (1993) based on socioeconomic data. A pyramid-like distribution of indications is proposed for radiotherapy with different beam modalities. According to availability this distinguishes between different stages during the introduction of ion beam therapy.

It is concluded by the author that in the first stage of clinical availability 64% of all patients who receive radiotherapy will be selected for conventional photon/electron therapy, 25% of patients could be treated by optimized fast neutron therapy, 7% will be treated with protons while 4% of patients will present with a indications for heavy charged particle treatment.

Results of Clinical Application of Ion Beam Radiotherapy

Uveal Melanoma

Posterior uveal melanoma has traditionally been treated by enucleation of the affected eyeball. However, some data have suggested that the prognosis after enucleation may be worse than for an untreated patient (Zimmermann and McLean 1975). Radiation therapy has been widely employed in the treatment of choroidal melanoma with a variety of techniques. When similar populations are compared, radiation therapy produces similar survival rates to enucleation. The treatment of uveal melanoma with proton and helium irradiation has been studied for more than 15 years. Particle therapy has been reported to yield higher local control rates than common brachytherapy techniques (Char et al. 1993). The results from different institutions are listed in Table 2.

Patients have been treated at various dose levels from 50 to 80 GyE in 4 to 10 fractions. The local control rate was 98%. Retention of the eyeball was possible in 85%, and in 50% of the patients useful vision could be preserved. The visual outcome of the patient was judged good, despite the exceptionally high doses employed. The visual result was highly sensitive to the proximity of the tumor to the fovea and the optic disk. Regarding the excellent tumor control rates, the doses prescribed were reduced subsequently and side effects could be minimized. Thirteen of 208 patients with large tumors at high dose levels required enucleation for complications. The leading side effect was anterior chamber complication (Meecham et al. 1994). Ninety percent of patients treated for ocular melanoma have avoided the need for surgical removal of the eye and have retained useful vision.

Table 2. Results of ion beam therapy of uveal melanoma

Author	No. of patients	Local control (%)	Survival (%)
Munzenrider (1994)	1957	96	80
Castro et al. (1994)	347	98	80
Egger and Zografos (1994)	1229	92	88

A randomized trial in uveal melanoma revealed a significantly higher tumor control with ion beam treatment than with ^{125}I-brachytherapy (100% vs 87%) (Char et al. 1993). More anterior segment complications occurred after helium ion irradiation. However, enucleation was more frequent after brachytherapy.

About 20% of all patients treated for ocular melanoma have occult metastases at the time of presentation. The risk factors for distant metastases are tumor diameter, anterior location and epitheloid cell histology. The prognosis of metastatic ocular melanoma is poor. Median survival times range between 2 months for untreated patients and 5 months for treatment of the metastases.

Tumor of the Skull Base

Tumors of the skull base are rare. The treatment of these tumors is associated with many problems, due to the adjacent structures. The immediately adjacent brain, brain stem, cranial nerves, blood vessels, eyes, etc., make definitive surgical excision often not feasible. Similarly these structure are radiation sensitive. The acceptable doses to these structures are below the dose for sufficient tumor control in skull base sarcomas. Tumors of the skull base can have very different histological features: chordoma, chondrosarcoma, meningeoma, and neurofibroma. However, there are also tumors of the nasopharynx extending into the skull base. There is only limited literature on photon treatment of sarcomas of the skull base. Local tumor control has been estimated to be 36% at 3.5 years (Austin-Seymour et al. 1985). With conformal treatment the dose in the target volume can be escalated without severe radiation-induced morbidity to the adjacent structures. There are only limited data on the treatment results of modern conformal treatment techniques. The clinical data of proton treatment have confirmed the theoretical considerations regarding higher tumor control probability. Reported tumor control rates by proton treatment are 91% for chondrosarcomas and 65% for chordomas of the skull base (Suit 1992). No severe morbidity has been reported in these patients. Castro reports Kaplan-Meier 5-year local control rates of 85% for meningeoma, 78% for chondrosarcoma, 63% for chordoma and 58% for other sarcomas (Castro et al. 1994). The mean dose was 65 Gy equivalent.

Soft Tissue Sarcoma and Bone Sarcoma

Patients with unfavorable soft tissue sarcoma underwent light ion treatment at the Lawrence Berkley Laboratory (LBL) in a phase I/II study. Especially patients with retroperitoneal sarcoma did notably well. The local control rate after 3 years was 64% with no radiation-related deaths. These results seem promising compared to historical data achieved by low-LET radiotherapy (Linstadt et al. 1991).

Seventeen patients with unfavorable bone sarcoma were treated wholly or in part with heavy charged particle irradiation at the LBL (Uhl et al. 1992). The majority of the tumors were located near critical structures such as the spinal cord or the brain stem. The 5-year Kaplan-Meier local control rate was 48%. However, more than half of the patients developed distant metastases despite the majority of the patients receiving chemotherapy.

Bile Duct Carcinoma

Schönthaler et al. found in a retrospective study that patients treated with helium or neon ions at a medium dose of 60 GyE had a significantly better survival (44%) than patients who received 54 Gy of conventional photon therapy (28%). There was also an advantage in the disease-free survival for the particle irradiation-treated patients, although this has to be verified in a randomized trial.

Other Indications

Proton beam treatment has been successful against tumors at other sites. For example, phase III trials of T3 carcinoma of the prostate indicate improved local control rates. The results of ion beam therapy in Bragg-peak radiosurgery have not been included in this review. Much experience has been gained in the ion treatment of arteriovenous malformations at BL and the Harvard Cyclotron.

Currently worldwide there are only two heavy ion facilities operational, both using carbon ion beams. The largest experience has been gained at the facility at Chiba, Japan, which started patient treatment in 1994. More than 300 patients have been treated in controlled nonrandomized clinical trials. The authors were able to demonstrate excellent clinical results in head and neck carcinomas compared to historical controls. The heavy ion facility at the Gesellschaft für Schwerionenforschung (GSI) in Germany started patient treatment in 1997. The clinical trials focus on the treatment of the base of skull tumors.

Discussion

The physical limits to gains from reductions of the treatment volume are imposed by the physical characteristics of the radiation beam and the errors involved in the treatment planning and execution process. The planning target volume has to take into account patient movement and setup errors. The absolute limit is reached when the treatment volume reaches the planning target volume ("if the treatment volume fits the target volume like a swimming-suit"). This is called conformal radiotherapy, where the amount of treated nontarget tissue is minimized. Clinical errors include the uncertainties in the definition of the target volume, etc.

All these factors will bring uncertainty to the clinical application of conformal techniques. Therefore it will be difficult to demonstrate a gain in tumor control probability with the use of an irradiation technique alone. As the target definition becomes the crucial point in the treatment process, rules for target definition in tree-dimensional-treatment planning need to be considered.

Is ion beam application:

1. Feasible in clinical routine? The feasibility of heavy ion treatment can now be demonstrated worldwide with more than 20000 patients. Therefore the feasibility of ion beam application in clinical routine has been clearly demonstrated. Modern feasibility studies will concentrate on technical improvements such as scanning beam techniques and gantry techniques.
2. Affordable? It is very critical from the ethical point of view to look at the economic aspects and the cost of curative treatment. The cost has to be compared with that of alternative treatments. Cost analysis is required to ensure that a routinely running heavy ion beam treatment facility will be cost effective compared to alternative treatment modalities [9]. However, the question is what is affordable. This depends on the economic status of the country. It should also be kept in mind that a relapse of a patient with a slowly progressing disease may be far more expensive due to health care costs than any currently available therapy.
3. Effective in the treatment of cancer? The evaluation of ion beam therapy in the treatment of cancer is difficult. The question of whether ion beam treatment is superior to conventional treatment has not been answered in general but needs further investigation. The two principal aspects to designing clinical tests of ion beam therapy are the definition of the reference treatment and the selection of appropriate patients.

Clearly the control arm has to be designed to be as close as feasible to the optimal photon/electron treatment methods in use. This means it is not appropriate to compare sophisticated ion beam techniques with conventional photon techniques. It is therefore necessary to design a reference treatment which is based on the recent developments in photon treatment such as three-dimensional treatment planning with multileaf collimators. The same technique of target definition and visualization and the same immobilization technique. If the design of a study does not fulfill these criteria, the conclusion of the trial could only be that advanced treatment techniques may be better.

Suit (1992) proposed criteria for patient selection for randomized trials regarding the application of ion beam therapy:

1. An advantage in local control can only be demonstrated if the equivalent dose in the target is 10% higher than in the reference treatment and the tumor control probability of the reference treatment is in the mid range ($0.2 < P < 0.6$).
2. An advantage regarding treatment morbidity can only be demonstrated if an important level of radiation-related morbidity of nontarget tissue is expected in the reference treatment.

According to these criteria, tumors of interest include: carcinomas of the head and neck (nasopharynx, maxillar and ethmoid sinus, salivary gland), prostate, uterus, and rectum, soft tissue sarcomas, gliomas, meningeomas, retroperitoneal tumors, and tumors in pediatric age patients. Especially in pediatric patients, where cure rates are high, the reduction of treatment-related morbidity is becoming more and more important. Therefore international participation in clinical studies is necessary to investigate the effectiveness of heavy ion treatment in phase III studies. Due to the limited availability of ion beam therapy, clinical studies will need to run for a long time. Another reason is that a sufficiently long-term follow-up is needed to completely evaluate a curative treatment.

References

Austin-Seymour M, Munzenrider JE, Goitein M, Gentr R, Gragoudas E, Köhler P, McNulty P, Osborne B, Ryugo DK, Seddon J, Urie M, Verhey L, Suit H (1985) Progress in low-LET heavy particle therapy: intracranial and paracranial tumors and uveal melanomas. Radiat Res 104:S219–S226

Castro JR, Linstadt DB, Bahary JP, Petti PL, Daftari I, Collier JM, Gutin PH, Gauger GE, Philipps TL (1994) The experience in charged particle irradiation of tumors of the skull base: 1977–1992. Int J Radiat Oncol Biol Phys 23:647–655

Char DH, Quivey JM, Castro JR, Kroll S, Phillips T (1993) Helium ions versus iodine 125 brachytherapy in the management of uveal melanoma. A prospective, randomized, dynamically balanced trial. Ophthalmology 100:1547–1554

Debus J, Hug EB, Liebsch NJ, O'Farrel D, Finkenstein D, Efird J, Munzenrider LE (1997) Brainstem tolerance to conformal radiotherapy of skull base tumors. Int J Radiat Oncol Biol Phys 39(5):967–975

Drumm K (1993) Sozioökonomische Studie zur Strahlentherapie mit geladenen Partikeln. Thesis, Rupprecht-Karls University, Heidelberg

Egger E, Zografos L (1994) Proton beam irradiation of uveal melanomas at PSI: latest results. In: Amaldi U, Larson B (eds) Hadron therapy in oncology. Elsevier, Amsterdam, pp 145–145

Engenhart R, Kimmig BN, Hover KH, Wowra B, Romahn J, Lorenz WJ, van Kàick G, Wannenmacher M (1993) Long-term follow-up for brain metastases treated by percutaneous stereotactic single high-dose irradiation. Cancer 71:1353–1361

Field SB, Fowler JF (1984) The biological basis for neutron therapy. J Eur Radiother 5:170–175

Gademann G, Wannenmacher M (1992) Charged particle therapy to pediatric tumors of the retroperitoneal region: a possible indication. Int J Radiat Oncol Biol Phys 22:375–381

Goitein M (1992) 3D treatment planning for heavy charged particles. Radiat Environ Biophys 31:241–245

Griffin TW, Krall JM, Laramore GB, Parker RG, Burnison M, Peters LJ, Maor M, Zink S, Davis LW, Russell KJ (1993) Fast neutron irradiation of prostate cancer: the results of a co-operative, phase III randomized study (Abstr). Int J Radiat Oncol Biol Phys 27 Suppl 1:247

Griffin TW, Wambersie A, Laramore G, Castro J (1994) High LET: heavy particle trials. Int J Radiat Oncol Biol Phys 14 S1: S83–S93

Hall EJ, Kraljevic J (1976) Repair of potentially lethal radiation damage: comparison of neutron and X-ray RBE-implications for radiation therapy. Radiology 121:731

Linstadt DB, Castro JR, Phillips TL (1991) Neon ion radiotherapy: results of the phase I/II clinical trial. Int J Radiat Oncol Biol Phys 20:761–769

Meecham WJ, Char DH, Kroll S, Castro JR, Blakely EA (1994) Anterior segment complications after helium ion radiation therapy for uveal melanoma. Radiation cataract. Arch Ophthalmol 112:197–203

Munzenrider JE (1994) Proton therapy with the Harvard cyclotron. In: Amaldi U, Larson B (eds) Hadron therapy in oncology. Elsevier, Amsterdam, pp 83–101

Schoenthaler R, Castro JR, Halberg FE, Phillips TL (1993) Definitive postoperative irradiation of bile duct carcinoma with charged particles and/or photos. Int J Radiat Oncol Biol Phys 27:75–82

Suit HD (1988) The scope of the problem of primary tumor control. Cancer 61:2141–2147

Suit H (1992) Potential for proton beams in clinical radiation oncology. In: Dewey WC, Edington M, Fry RJM, Hall BJ, Whitmore GF (eds) Radiation research. A twentieth century perspective, vol 2. Academic, San Diego, pp 3–13

Uhl V, Castro JR, Knopf K, Phillips TL, Collier JM, Petti PL, Daftari I (1992) Preliminary results in heavy charged particle irradiation of bone sarcoma. Int J Radiat Oncol Biol Phys 24: 755–759

Zimmermann LE, McLean IW (1975) Changing concepts concerning the malignancy of ocular tumors. Arch Opthalmol 78:487

Boron Neutron Capture Therapy: Principles and Potential

R. Gahbauer[1], N. Gupta[1], T. Blue[2], J. Goodman[3], R. Barth[4], J. Grecula[1], A.H. Soloway[5], W. Sauerwein[6] and A. Wambersie[7]

[1] Division of Radiation Oncology, Ohio State University, Columbus, OH 43210, USA
[2] Nuclear Engineering Program, Ohio State University, Columbus, OH 43210, USA
[3] Division of Neurosurgery, Ohio State University, Columbus, OH 43210, USA
[4] Department of Pathology, The Ohio State University, Columbus, OH 43210, USA
[5] Department of Medicinal Chemistry, Ohio State University, Columbus, OH 43210, USA
[6] Department of Radiotherapy, University of Essen, Hufelandstrasse 55, D-45122 Essen, Germany
[7] Université Catholique de Louvain, Cliniques Universitaires St-Luc, Avenue Hippocrate 54, B-1200 Brussels, Belgium

Abstract

This book on the therapeutic applications of neutrons and high-LET radiations in cancer therapy would not have been complete without a review of the present situation of boron neutron capture therapy (BNCT) and a discussion of its future perspectives. BNCT is a special type of high-LET ratiation therapy that attempts to achieve a selectivity at the cellular level. The rationale is to incorporate boron atoms selectively in the cancer cells and then bombard those atoms with thermal neutrons to produce a neutron capture reaction and subsequent decay that emits alpha and lithium particles. The efficiency of the technique depends upon achieving selective incorporation of the boron atoms in the cancer cells and not (or to a lesser extent) in the normal cells. The present status and future directions are described, with emphasis on boron carriers (drugs) and their delivery, as well as physical and treatment planning aspects.

Introduction

The concept of BNCT was introduced by Locher in 1936, shortly after the discovery of the neutron (Slatkin 1991; Barth et al. 1996). Farr at Brookhaven National Laboratory (BNL) and Sweet at Massachusetts General Hospital/ Massachusetts Institute of Technology (MGH/MIT) carried out the first clinical trials during the 1950s and early 1960s. Suboptimal neutron beams and poor boron compounds led to their failure. In Japan, Hatanaka and Nakagawa (1994) resumed clinical trials in 1968. The results he reported rekindled the interest in BNCT worldwide. The rationale for BNCT is attractive and could open new perspectives in radiation therapy. For physical and biological

Recent Results in Cancer Research, Vol. 150

reasons, its main benefit may be shown in glioblastoma multiforme, a tumor for which present treatments are unsatisfactory. For these reasons, the Department of Energy has supported research and therapy trials in United States. Since September 1994, more than 40 patients have been treated at BNL and MIT. Likewise the European Commission has supported a multinational BNCT project in Petten (Netherlands). The first patients were treated there in October 1997.

A quite different mode of BNCT was proposed in 1978 by Waterman et al. and is still being explored for potential therapeutic applications in Essen, Seattle, and Detroit. In this concept, the boron neutron capture reaction is used to augment fast neutron therapy (Sauerwein et al. 1989; Laramore et al. 1994).

In the treatment of malignant brain tumors, the use of conventional, low-LET radiation has been shown to increase median survival 2.5-fold (Walker et al. 1978); however, permanent tumor control is very rarely achieved even at very high doses (Salazar et al. 1979). As in all other attempts to find a therapeutic window for glioblastoma multiforme (Fig. 1), fast neutron trials failed to achieve tumor control within the tolerance limits of normal tissues. However, in most reports from these trials, it was suggested that a better tu-

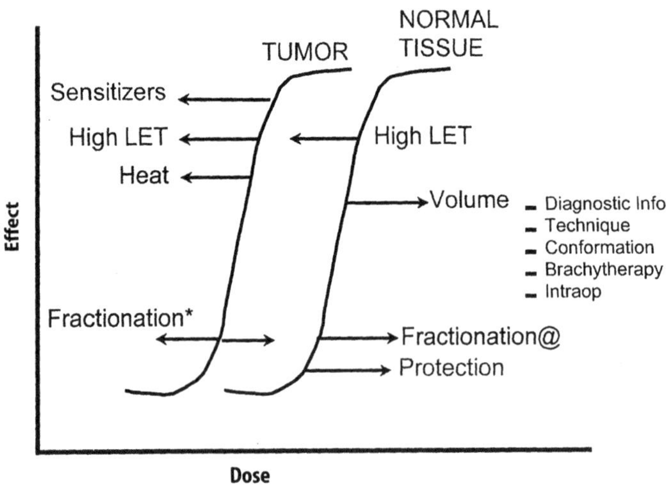

@ Fractionation can protect normal tissues by allowing repair of sublethal (and potentially lethal?) damage and allowing some proliferation

* Fractionation also protects tumor cells by allowing repair of sublethal (and potentially lethal?) damage, and allowing malignant cell proliferation. In contrast, fractionation may increase the radiosensitivity of the tumor, e.g. eliminating hypoxic cells ("reoxygenation")

Fig. 1. Sparing of normal tissues, and preferential or selective effects on tumor cells vs normal tissues obtained by different treatment techniques

moricidal effect was observed than that seen with equivalent high doses of low-LET radiation. Laramore et al. concluded in 1988 that clinical trials could only resume if it were possible to deliver high-LET radiation to tumor cells with greater selectivity (e.g., using such techniques as monoclonal antibodies). Conceptually, boron neutron capture therapy is an ideal method to selectively deliver high-LET radiation to tumor cells. Beyond the intrinsic compound dependent, cellular specificity of BNCT, the capture reaction can be restricted to a limited target volume because it requires activation by thermal neutrons.

Principles

BNCT is based on the nuclear fission that occurs when nonradioactive ^{10}B (in nature 20% of elemental boron) absorbs thermal neutrons (0.025 eV), as shown in Fig. 2. Alpha particles and lithium nuclei with a range of 4–7 µm emanate from this reaction, sharing between them 2.3–2.8 MeV. The alpha particle produces the main radiation effect due to its longer range compared to that of the ^{7}Li nucleus. The gamma radiation produced contributes very little to the local or normal tissue effect. The radiation selectivity of BNCT derives from the fact that the capture cross-section values of normal tissue elements for thermal neutrons are two orders of magnitude less than of boron-10. The thermal neutron capture cross-sections of a few capture agents of interest and of components of normal tissue are listed in Tables 1 and 2. Of the elements present in normal tissue, only hydrogen (gammas) and nitrogen (protons) contribute significantly to the dose delivered to normal tissues and tumor by capture reactions $[^{1}H(n,\gamma)^{2}H]$ and $[^{14}N(n,p)^{14}C]$. The energy loss of epithermal neutrons in tissue by scattering with hydrogen generates energetic recoil protons $[^{1}H(n,n')^{1}H]$ that deposit significant radiation dose in tissues. Gamma radiation emanating from the reactor or accelerator also must be considered.

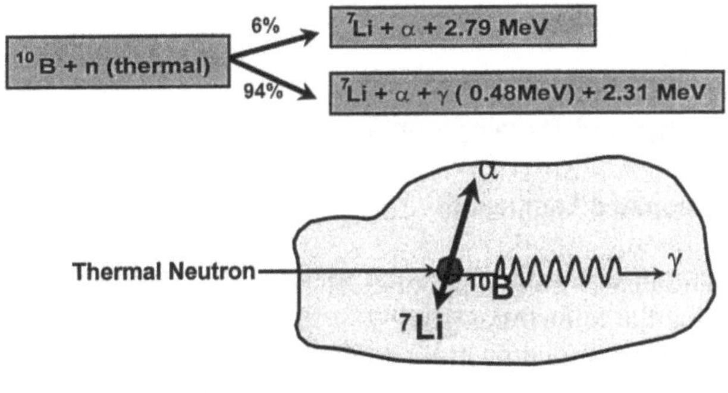

Fig. 2. The $^{10}B[n,\alpha]^{7}Li$ reaction, which is the bases of boron neutron capture therapy. Schematic sketch of the reaction below shows the relative ranges of the reaction products

Table 1. Thermal neutron capture cross-sections (barns) for some capture agents

Nuclide	Neutron capture cross-section values	Nuclide	Neutron capture cross-section values
^{6}Li	942	^{157}Gd	225 000
^{10}B	3 838	^{164}Dy	1 800
^{113}Cd	19 800	^{168}Yb	3 500
^{149}Sm	42 000	^{184}Os	3 000
^{151}Eu	5 800	^{196}Hg	3 000
^{155}Gd	61 000	^{199}Hg	2 000

Table 2. Composition of tissue and thermal neutron capture cross-sections (barns) of each component

Nuclide	Weight % in tissue	Neutron capture cross-section	Nuclide	Weight % in tissue	Neutron capture cross-section
H	10.0	0.332	P	1.16	0.18
C	18.0	0.0034	S	0.2	0.53
N	3.0	1.82	Cl	0.16	32.68
O	65.0	1.8×10^{-4}	K	0.2	2.1
Na	0.11	0.43	Ca	2.01	0.4
Mg	0.04	0.053	Fe	0.01	2.57

To meet the criteria of selectivity and effectiveness, there must be a significantly higher boron concentration in tumor than in normal tissues and approximately 20–40 μg/g B in tumor (10^{9} ^{10}B atoms/cell). There are a number of nuclides with even higher cross-sections than boron, some of which have been evaluated as possible capture agents, e.g., gadolinium (Mill et al. 1994; Matsumoto 1993). The main interest, however, is focused on boron, because the range of the high-LET particle is limited to approximately 1 cell diameter, resulting in a very local (selective) energy deposition, and because boron compounds rival carbon in their extensive covalent chemistry and stability, allowing for the synthesis of diverse chemical entities from low to macromolecular species.

Status and Developments

Boronated Compounds

The ideal boron compound for targeting malignant brain tumors should possess the following attributes:
(1) it must demonstrate an ability to target tumor cells and their microscopic extensions selectively, even those that may be protected by the normal blood-brain barrier (BBB);

(2) it should achieve at least on the surface but preferably in an intracellular location within all tumor cells a concentration level of 10^9 boron-10 atoms/cell when administered alone or in combination with other boron compounds (the ideal site would be the tumor DNA, since this is the purported target of the high-LET radiation);
(3) tumor to brain and blood ratios in excess of 3–4/1 would be highly desirable;
(4) compound persistence especially during the course of the irradiating procedure is essential; and
(5) the compound or compounds must be sufficiently nontoxic systemically so that adequate tumor concentrations by intravenous or intracarotid administration can be attained.

One objective of modern compound development is to design low molecular weight agents related to natural biochemical constituents, based on the rationale that cycling cells (tumor cells vs non cycling normal brain cells) have a greater requirement for "building blocks." Among the boron compounds that have been developed are phospholipid ethers that may selectively accumulate in plasma membranes (Lemmen et al. 1993), amino acids that may be incorporated into tumor proteins (Wyzlic et al. 1994), and nucleosides that can be converted into nucleotides to be incorporated into DNA or RNA (Rong and Soloway 1994).

Other approaches have involved the devlopment of porphyrins (Kahl and Koo 1992), DNA intercalators (Tjarks et al. 1994) radiation sensitizers (Scobie and Threadgill 1992) and polyamines (Hariharan et al. 1995). Progress has been made with macromolecular species with the synthesis of boronated liposomes (Feakes et al. 1993), low density lipoproteins (Kahl et al. 1993), antisense oligonucleotides (Spielvogel et al. 1993), monoclonal antibodies and their fragments, bispecific antibodies (Liu et al. 1995), and tumor receptor binder such as epidermal growth factor (Capala et al. 1996). Disruption of the BBB is being investigated for those compounds that are otherwise unable to reach microscopic extensions of tumor (Yang et al. 1996).

Boron Compound Delivery: Optimizing Delivery of Boron-Containing Agents to Brain Tumors

Considerable effort has been directed towards the design and synthesis of low and high molecular weight boron delivery agents to target brain tumors. Until recently, however, little effort has been directed towards developing strategies to maximize their uptake by tumors and to concomitantly minimize normal brain and blood levels. The intravenous route currently is being used clinically to deliver both BSH and BPA, and, although this is convenient, it may not be ideally suited for delivering high and low molecular weight agents to brain tumors. Strategies for enhancing the delivery of drugs to brain tumors broadly can be classified as invasive, pharmacologic, or physiologic (Pardridge 1995, 1997; Riva et al. 1994). Invasive techniques include direct intratumoral injection (Riva et al. 1994), the implantation of sustained release polymers (Rapoport 1980; Tamargo et al. 1994), convection enhanced delivery (Lieberman et al. 1995), and hyperosmotic (1.373 mosmol) mannitol-induced disruption of the BBB (Neuwelt and Frenkel 1989; Gumerlock and Neuwelt 1993). Pharmacologic approaches include the use of small, lipid soluble molecules and liposomes with diameters <50 nm (which will traverse the BBB), and the bradykinin agonist, RMP-7 a synthetic monopeptide that has been reported to selectively open the BBB within brain tumors (Matsukado et al. 1996). Physiologic strategies include the use of pseudonutrients such as insulin like growth factor-1 (IGF-1), cationic antibodies, and chimeric peptides (Pardridge 1997).

The BBB normally prevents the passage of ionized, water soluble drugs that have molecular weights >180 Da. Even if the main bulk of the brain tumor has a permeable vascular endothelium, the brain around tumor, which contains invasive tumor cells, has decreased permeability to antitumor agents. Endothelial cell clefts in microvessels of the tumor may vary widely from normal to abnormal within different regions of the same tumor, indicating the complexity of the tumor vasculature. Neuwelt and Frenkel (1989; Gumerlock and Neuwelt 1993) have strongly advocated the use of hyperosmotic mannitol-induced blood brain barrier disruption (BBB-D) to enhance the delivery of cytoreductive chemotherapeutic agents in patients with brain tumors. Their most recent data suggest that this approach can significantly enhance the survival of patients with primary central nervous system lymphomas (Dahlborg et al. 1996). However, on the negative side, BBB-D combined with the intra-arterial administration of cytoreductive chemotherapeutic agents has been associated with increased toxicity (Shapiro et al. 1988). Pharmacodynamic studies have shown that agents with rapid blood clearance, moderate lipophilicity, and low neurotoxicity are more effectively delivered by the intra-arterial route (Fenstermacher and Cowles 1977). This has important clinical implications for both BPA and BSH, which currently are being used in clinical trials of BNCT for patients with glioblastomas.

Enhancing the delivery of either BPA or BSH can have a dramatic effect both on increasing brain tumor boron uptake and therapeutic efficacy. Using the well characterized F98 rat glioma model (Barth 1998), intracarotid (i.c.) injection of either BPA or BSH doubled boron uptake in tumor; this could be increased fourfold by BBB-D, compared to uptake obtainable by intravenous (i.v.) injection. In contrast, normal brain boron values at 2.5 h after injection were very similar for i.v. and i.c. delivery or for i.c. in combination with BBB-D (Barth et al. 1997a). Since BNCT is a binary system, normal brain boron levels only are of significance at the time of irradiation, and high values at earlier time points would be inconsequential.

BNCT studies were carried out by us at the Brookhaven National Laboratory to compare the therapeutic efficacy of i.v. versus i.c. administration of BPA or BSH with or without BBB-D using the F98 rat glioma model. Survival times of animals that received BPA or BSH i.c. with BBB-D were increased 188% and 117%, respectively, compared to that of irradiated controls. The best survival data were obtained using both compounds in combination with i.c. injection and BBB-D, and 25% F98 glioma bearing rats were cured of their tumors for the first time ever (Barth et al. 1997b). These results convincingly demonstrated that a significant therapeutic gain could be obtained by optimizing boron compound delivery; this has important implications for both the ongoing clinical trials with BPA and BSH, and those that may be initiated in the future.

Different strategies may be required for other low molecular weight boron containing compounds whose uptake is cell cycle dependent or for high molecular weight agents such as monoclonal or bispecific antibodies or boronated proteins such as epidermal growth factor (EGF). For example, we recently reported that direct intratumoral delivery of boronated EGF was both selective and effective in delivering therapeutically effective amounts of boron to the tumors of glioma bearing rats compared to the extremely small quantities that were delivered by the i.v. route (Faillot et al. 1996). In the future, optimizing of boron compound delivery will be an integral part of the pre-clinical and clinical evaluation of new capture agents for BNCT.

BNCT Neutron Sources

It has been recommended that for BNCT to be successful, a thermal neutron fluence of about 5×10^{12} n/cm^2 be delivered to a tumor (Barth et al. 1990) with a ^{10}B concentration of 30 μg/g. Early efforts in BNCT research (Sweet 1983) used thermal neutron beams (neutron energies <1 eV) to treat patients with deep seated tumors in clinical trials. Due to the short mean free path of thermal neutrons, the flux levels for deep seated tumors were insufficient to provide the total fluences necessary for successful treatment without excessive damage to normal tissue; and the clinical trials were a failure. It was then decided that epithermal neutrons (1 eV<neutron energies <10 keV) were necessary for the treatment of deep seated (≥2 cm deep) tumors. Epithermal neutrons thermalize at a depth of about 2.5 cm. They provide a maximum thermal neutron fluence at the tumor site, with a minimum of damage to normal tissue.

Reactor Neutron Sources

Initially, only reactors were thought to be capable of delivering the necessary fluence of neutrons in a suitably short time. The first clinical trials of BNCT were carried out at the nuclear reactors at the Brookhaven National Laboratory (BNL) and the Massachusetts Institute of Technology (MIT) – both using thermal neutron beams. Subsequently, in Japan, BNCT treatments were carried out at the Hitachi Training Reactor (HTR), the Japan Atomic Energy Research Institute Research Reactor (JRR), the Musashi Institute of Technology Reactor, and the Kyoto University Reactor (KUR). All these reactors provided thermal neutron beams for BNCT treatments, but only the JRR and the KUR are currently available for BNCT treatments. The KUR staff has plans to remodel their filter design to provide an epithermal beam in addition to the currently available thermal neutron beam (Kobayashi et al. 1993). The team involved in European Collaboration for BNCT have modified the high flux reactor (HFR) at Petten in the Netherlands to obtain an epithermal beam (Moss 1990) for their animal studies and have recently begun human clinical trials.

In the United States, two reactors, the Brookhaven Medical Research Reactor (BMRR) and the Massachusetts Institute of Technology Reactor (MITR), are currently being used for BNCT clinical trials with epithermal beams. Prompted by the advent of clinical trials, other research reactors, such as the Georgia Institute of Technology Research Reactor (GITRR) and the Power Burst Reactor Facility (PBF) at Idaho Falls, Idaho, were also being evaluated for BNCT. In addition, the staff of a number of research reactors around the United States are considering upgrading their facilities for possible BNCT use.

The possible use of fission plates to improve the epithermal neutron flux at a relatively low operating power is being actively considered at a number of research reactors (Liu et al. 1994; Harling et al. 1996a). Fission plates contain subcritical quantities of a fissile material. A fission plate converter design places the fission plate between the reactor core and the treatment port to achieve subcritical multiplication of the neutron flux and a resultant higher neutron flux. Figure 3 shows a conceptual design for a fission plate converter being studied at The Ohio State University (OSU) Research Reactor.

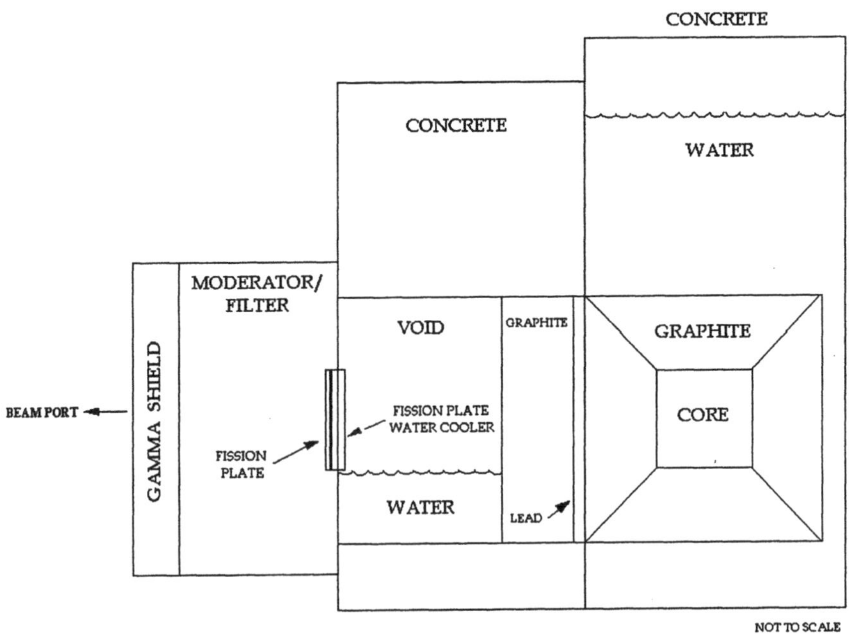

Fig. 3. A conceptual design of a fission plate converter being studied at the Ohio State University Research Reactor

Accelerator-Based Neutron Sources

Since it is highly unlikely that reactors can be situated in populated areas as a part of major medical centers, alternative sources of thermal and epithermal neutrons for BNCT are being investigated. The most appealing of the alternatives that has been suggested is the use of low-energy proton accelerators with low z targets (Blue et al. 1985). Among possible reactions studied were the $^{7}Li(p,n)^{7}Be$ reaction (Blue et al. 1985; Wang et al. 1989), and the $^{9}Be(p,n)^{9}B$ reactions (Wang and Moore 1994; Howard et al. 1996). Of these, the lithium target has been studied more extensively, because of the higher neutron yield obtained with the proton reaction on lithium, which yields neutrons of relatively lower energies that can be moderated/filtered easily. Figure 4 shows a conceptual view of an accelerator-based neutron source. Figure 5 is a configuration of a moderator assembly currently being tested at OSU. Accelerators are currently in use in radiation therapy and are well accepted in the medical community. With the advancement of technology, cur-

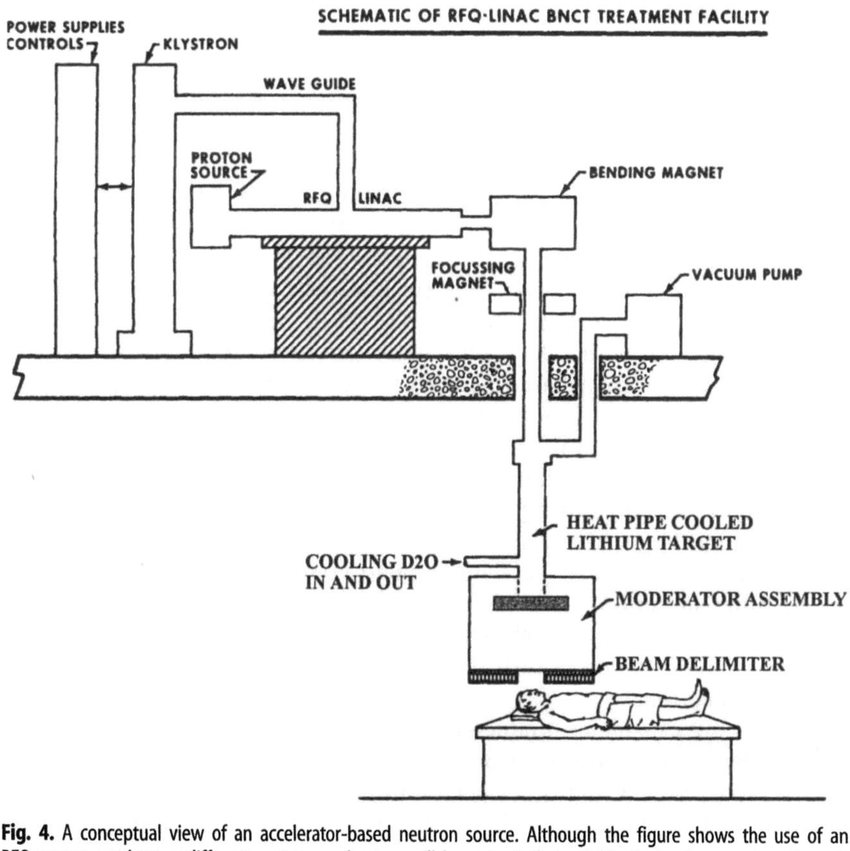

Fig. 4. A conceptual view of an accelerator-based neutron source. Although the figure shows the use of an RFQ proton accelerator, different proton accelerators will be potentially available for implementation of accelerator-based neutron sources for BNCT

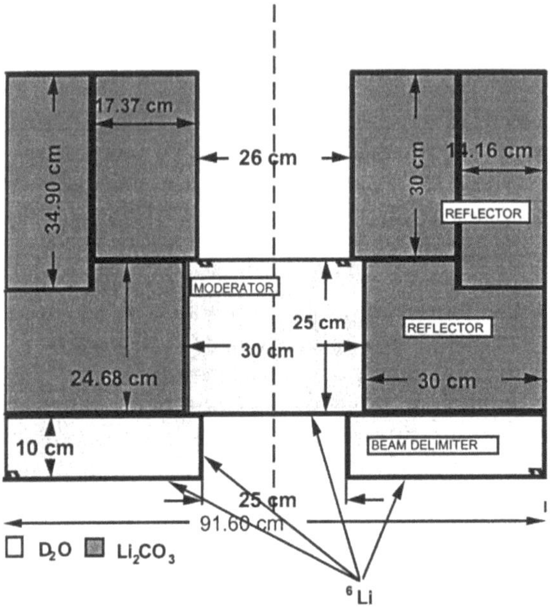

Fig. 5. Configuration of a fabricated moderator assembly being testes at Ohio State University for an accelerator-based neutron source for BNCT

rent accelerator designs are very compact compared to those of accelerators built a few decades back. Among the different types of proton accelerators proposed are Radio-Frequency Quadrupole (RFQ) linacs (Wangler et al. 1989; Hamm and Shubaly 1995), Tandem Electrostatic Accelerators (Shefer et al. 1992, 1995), Electrostatic Quadrupole (ESQ) accelerators (Kwan et al. 1995), and an Rf Focused Drift Tube Linac (Swenson 1995). The main components of an accelerator-based neutron source for BNCT are a high current proton linac, a target to produce neutrons by a suitable (p, n) reaction (with its associated heat removal system), and a moderator assembly to filter the energetic neutrons born at the target to energies suitable for therapy.

Deuteron accelerators using the $^9Be(d, n)$ reaction with 2.6 MeV deuterons on a beryllium target have also been proposed for BNCT applications (Buchholz et al. 1994). Another accelerator neutron source design using a completely different concept is a photoneutron source (Nigg et al. 1995; Jones and Yoon 1992), which uses electron linear accelerators to produce a bremsstrahlung photon, which in turn undergoes a photoneutron (γ, n) interaction to produce neutrons.

Reactors are capable of producing higher flux levels at the patient than accelerator based sources of neutrons. Hence, a single BNCT treatment session with an accelerator would be longer. However, use of fractionated treatments would allow for retargeting of boron and repair of sublethal damage to normal tissue from the low-LET dose components, thereby increasing tolerance (Gahbauer et al. 1989, 1997) and making this disadvantage of the accelerator based neutron source less important. For example, it has been estimated that it may take 36 min to treat a patient in a single treatment session with an accelerator based neutron source (Blue et al. 1995). If the treatment were fractionated over four treatment sessions, then the treatment time per session would be a very reasonable 9 min.

The proton energy chosen for accelerator based neutron sources utilizing the $^7Li(p, n)^7Be$ reaction was 2.5 MeV. This energy was chosen because the above reaction had a threshold of 1.88 MeV, and the neutron yield peaked at around 2.3 MeV. Hence it was felt that 2.5 MeV protons would be appropriately slightly higher than the energy at which the yield peaked so that the actual yield was not dependent on slight variations in accelerator operating conditions, resulting in a more stable neutron production.

One of the major challenges with accelerator sources utilizing the $^7Li(p, n)^7Be$ reaction has been target cooling. A 2.5-MeV proton beam impinging on a lithium target and delivering a 10-mA proton beam (the minimum current required to produce an epithermal beam of acceptable intensity) deposits 25 kW heat on the target and must be removed to prevent melting and subsequent vaporization of the lithium target (melting point of 180 °C). Various methods of heat removal have been proposed for such targets (Christensen et al. 1991; Vafai and Wang 1992; Huang and Vafai 1994; Blackburn et al. 1997), but doing this is not a trivial task.

Due to the above problems with target heat removal, recently researchers have proposed and are investigating the use of the $^7Li(p, n)^7Be$ reaction at energies just above threshold energies (Harker and Harmon 1997). The rationale behind lowering the proton energy is that neutrons are generated at lower energies; therefore, thinner filters than those used with neutrons produced with 2.5 MeV protons on lithium can be used. Although the neutron yield is reduced considerably when the proton energy is lowered, the reduction of filter thickness produces comparable neutron fluxes at lower proton currents. The lower proton current coupled with the lower proton energy would mean less heat production at the target to deal with. However, operating a target at just above threshold energies would cause the neutron yield to be very dependent on the exact proton energy delivered to the target (due to the shape of the neutron yield vs proton energy curve), and even slight variations in the machine output would cause the neutron flux to fluctuate.

Other Alternative Neutron Sources

Besides the accelerator based neutron source, other alternative neutron sources proposed for BNCT are spallation neutron sources (Grusell et al. 1990), a californium-252 (Kim et al. 1993) neutron source, and an inertial electrostatic confinement (IEC) source (Miley et al. 1995). BNCT has also been suggested as a means of boosting the dose to the tumor in external beam fast neutron therapy (Laramore et al. 1994) and fast neutron brachytherapy (Beach et al. 1990). Among the proposed non-reactor neutron sources, accelerator based neutron sources have probably received the most attention.

Dosimetry, Evaluation of Radiation Quality and Treatment Planning

Dose Components in BNCT and Their Quality

The boron dose (dose from the $^{10}B[n,\alpha]^{7}Li$ reaction) and the nitrogen capture dose (henceforth called *nitrogen dose*) inside the head have the same spatial dependence as the thermal neutron fluence distribution, since these interactions have a huge cross-section (interaction probability) for thermal neutrons and the resultant particles deposit energy locally. The hydrogen capture gamma dose also has the same spatial dependence as the thermal neutron fluence distribution, but the gammas do not deposit their energy locally; consequently, the gamma dose spatial distribution does not have the same spatial dependence as the thermal neutron fluence distribution. The dose from proton recoils (also called the *fast neutron dose*) depends on the neutron fluence energy spectrum and has roughly the same spatial dependence as the fast neutron fluence distribution. Figure 6 shows representative central axis absorbed dose distributions for the different dose components for a configuration of the Brookhaven Medical Research Reactor epithermal neutron beam.

The various dose components in BNCT represent a mix of high- and low-LET radiations. The mixed radiation field presents a unique problem in treatment planning in BNCT. The RBEs of the different components are not only different, they also change with depth in tissue (resulting from energy changes),

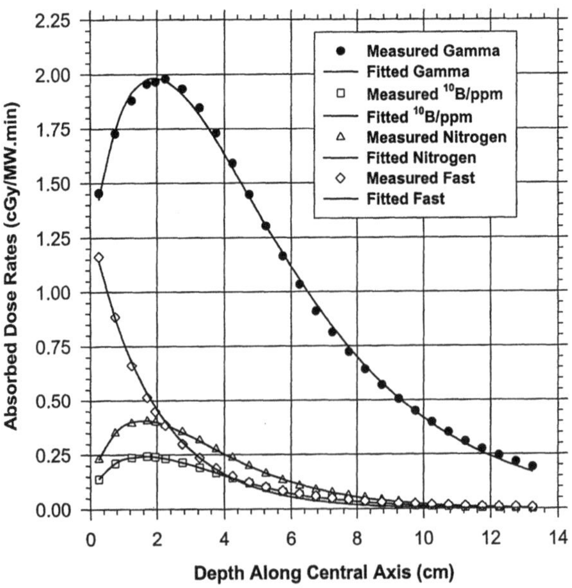

Fig. 6. Fits of measured central axis absorbed dose distributions for a configuration of the BMRR epithermal neutron beam. The data points show the measured doses, and the line through the data points are the fits to the data. (From: Floyd Wheeler, personal communication, April 1994)

as does the overall mix of high- and low-LET radiations (Gupta et al. 1994). The RBEs of the individual components cannot be determined directly, but are usually derived based on some simplifying assumptions from in vitro or in vivo experimental data or from data generated for various end points in small or large animal models. Depending on these simplifying assumptions, RBEs determined by different methods may vary considerably. In addition, the quality of each epithermal neutron beam to be used for BNCT is different, resulting in perceived differences in the RBEs of each beam's components, as well as necessitating reevaluation of each beam to generate the RBE data.

Dose from the $^{10}B[n, \alpha]^{7}Li$ Reaction in Normal Tissue and Tumor

Concept of Absorbed Dose. The quantity "absorbed dose" at a given point is used to specify the amount of the therapeutic irradiation at that point and is the relevant quantity commonly used for the different techniques, e.g., external beam therapy with photons, electrons, neutrons, protons, etc. The absorbed dose at a point is defined as the "mean energy imparted" per unit mass in a volume surrounding that point. The averaging process implies that a large number of ionizing particles are crossing the volume of interest (ICRU 1980). There is a strict relation between absorbed dose and cell lethality (e.g., the linear-quadratic relation). Because in BNCT different radiation components contribute to the absorbed dose at a given point, the situation is more complex and includes the following:

1) Gamma rays from the reactor and from the hydrogen capture
2) The "nitrogen dose"
3) Fast neutrons
4) The so-called "boron dose," which is the absorbed dose due to the alpha particles, resulting from the capture of thermal neutrons by boron atoms followed by the disintegration of the boron atoms

The concept of absorbed dose can be applied to the three first components, as in the conventional therapeutic radiation modalities; but it has limitations in the case of the boron dose, since the boron incorporation, and thus the number of alpha particles, is not uniform at the cellular level. It can (and probably does) vary from one cancer cell to another one (between zero and many). In such conditions, the "averaging process" implied in the definition of the concept of absorbed dose is no longer meaningful (Fig. 7).

Modifying Factors. In clinical situations, it is only possible to average the boron concentration over a large population of cells and thus derive a kind of average dose. If tumor cells do not contain boron or contain less boron than normal tissue, any increase in neutron fluence will yield no therapeutic benefit and will disproportionately increase the toxicity in normal tissues. Therefore, determining compound factors or modifying factors to establish a therapeutic gain calculation is irrelevant, since the paucity of boron in a given tumor cell population would cause a negative therapeutic gain with re-

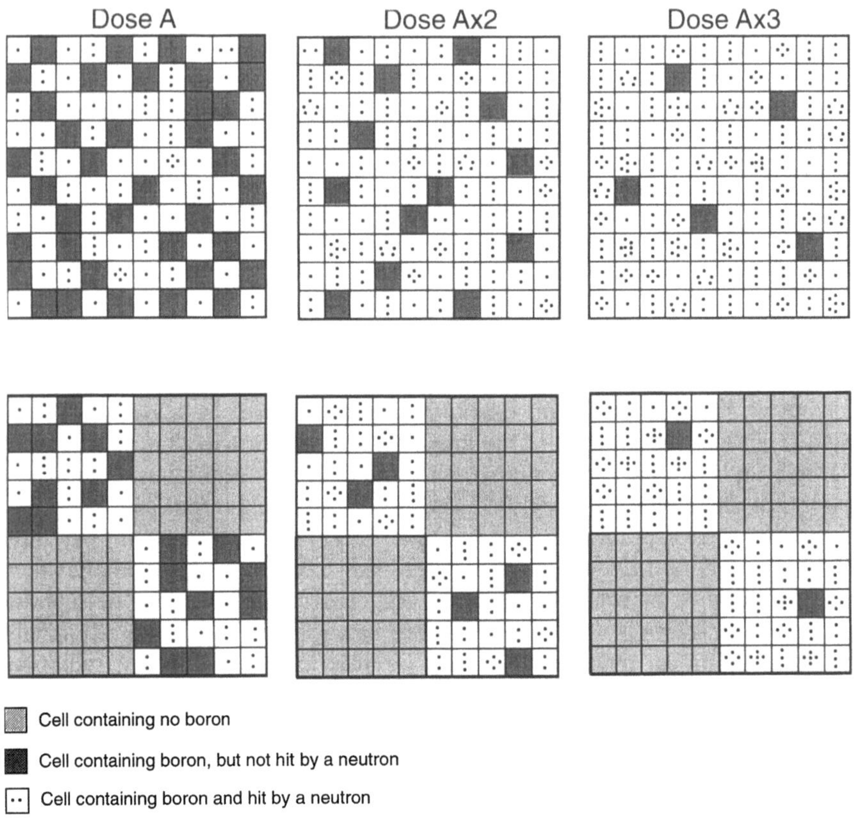

Fig. 7. The top panel illustrates a random distribution in 100 equal-sized "targets" of 100, 200, 300 "hits" and reflects the situation with conventional radiation. The probability that any one of the 100 targets will not be struck (i.e., will survive) when 100 hits are delivered randomly is e–l or 37% and e–3 or 5% after 300 hits. There is a strict relationship between the number of hits (dose) and survival (From Whiters and Peters 1980). The *bottom panel* illustrates the situation with respect to the boron dose. Here it is assumed that half the volume does not contain any boron. While the neutron dose is still averaged over the volume, cells without boron present no target and will not receive a strike, regardless of the number of hits (dose) delivered. Conversely, there is a possibility that some normal tissue may take up boron more avidly. (A non-boron dose is not considered in this example and the boron concentration is not increased.

spect to normal tissues, even if the average boron concentration indicates a significant gain (Fig. 8).

In normal tissues, the same difficulties apply to some degree. However, for a given patient, the boron concentration in blood can be determined accurately and repeatedly during radiation, and the distribution of boron may follow a relatively predictable pattern. Known RBEs and compound factors from experiments may be used to establish a starting point for phase I dose escalation studies. The ultimate tolerance must be determined by careful dose escalation studies in which particular attention must be paid to the possibility of unexpected boron uptake in a sensitive normal tissue population.

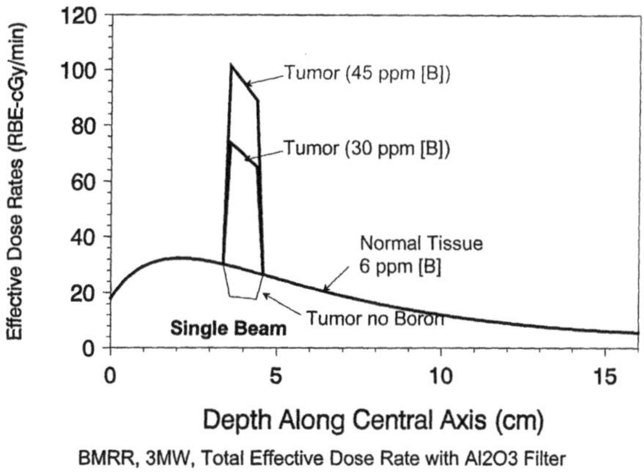

Fig. 8. Central axis total RBE-dose rate distributions with depth inside a head phantom with a single field treatment setup for a configuration of the BMRR epithermal neutron beam. The distributions were calculated assuming a uniform normal tissue boron concentration of 6 μg/g tissue, and with different uniform tumor boron concentrations for a 4-cm-deep tumor. If more boron is retained in tumor than in normal tissue, a large additional dose can be selectively delivered, suggesting a big therapeutic gain. However, if any part of the tumor contained less boron than normal tissues, a negative therapeutic gain would result. This negative therapeutic gain would be magnified by any escalation of the physical dose (i.e., increase in treatment time). Thus dose escalation studies in BNCT must aim at maximizing the number of cells containing boron as well as maximizing physical dose delivery.

Dose Escalation. The difficulty in determining the boron dose also reflects on the design of dose escalation studies in BNCT.

Dose escalation can be understood in three ways:

1. Escalation in exposure time
2. Escalation in the amount of boron administered or delivered to the tumor
3. Escalation in both factors

Escalation in exposure time will proportionally increase the contribution of the gamma dose (capture and reactor), the nitrogen and fast neutron dose contributing to the absorbed dose at the point(s) of interest. The normal tissues at risk will receive higher doses, and they will come closer to their tolerance limit.

As far as the boron dose is concerned, therapeutic gain can occur only when the boron concentration is higher (in principle) in all the cancer cells than in the normal tissues at risk. In contrast, if the boron concentration in some cancer cells is lower than that in the normal tissues at risk, increasing the exposure time will result in a negative therapeutic gain (this is true a fortiori in cases in which some cancer cells incorporated no boron at all).

The same arguments can be made when increasing the amount of boron per administration of the drug. In contrast, an improved therapeutic gain could be expected when the number of drug administrations is increased, since the boron distribution could vary from one administration to another and thus increase the probability of reaching all the cancer cells (e.g., alterations in blood supply and effect of previous irradiation(s)).

Compound Factor. The unknowns in localization of the ^{10}B compound and their dosimetric implications complicate the planning and evaluation of BNCT experiments and treatments. The knowledge of the localization of the ^{10}B compound in the normal brain and in the tumor are necessary for treatment planning (predicting normal tissue tolerances) and for predicting or modeling possible tumor response. A compound taken up intracellularly (by the cell cytoplasm or, better still, by the cell nucleus) will have a much higher biological effectiveness than a compound which remains in the extracellular spaces. Further, some compounds, like BSH, remain primarily in the intraluminal vascular spaces and provide an added "geometric protection" (Deutsch and Murray 1975; Rydin et al. 1976) to the capillary endothelium. This is illustrated in Fig. 9, where a boron reaction inside the capillary lumen is shown. The α or ^{7}Li particle can travel in any direction, and only a small fraction of these reach the nucleus of the endothelial cell, resulting in the "geometric protection."

From the macrodosimetric perspective, the unknowns in boron localization and the geometric protection have been combined into the so-called "compound factor" (Gupta et al. 1994; Gahbauer et al. 1992), which can be defined as the "RBE" of the boron dose component for a particular compound. The compound factor may be determined by small and large animal experiments with and without boron by comparing iso-effects.

Microdosimetry. Analytical and stochastic models for calculating the microdosimetric parameters based on various scenarios of boron subcellular distributions have been developed over the past decade, the most popular of these methods being Monte Carlo simulations (Gabel et al. 1987; Deutsch and Murray 1975; Rydin et al. 1976; Blue et al. 1993; Wheeler 1992; Kobayashi and Kanda 1993; Kalend et al. 1995). These models have been validated by microdosimetric measurement methods using tissue equivalent proportional counters (with boron-lined walls) to mimic various target sizes (Wuu et al. 1992). High-resolution autoradiographic methods that have been recently developed to measure subcellular boron uptake (Solares and Zamenhof 1995) will

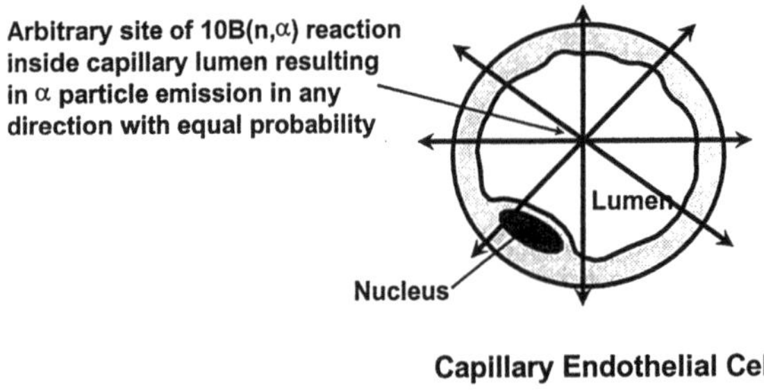

Fig. 9. The boron reaction with a thermal neutron inside the capillary lumen for compounds that remain inside the capillary lumen. The α or ^{7}Li particle can travel in any arbitrary direction, and only a small fraction of them reach the nucleus of the endothelial cell, resulting in a "geometric protection"

further validate these methods. These means of calculating microdosimetric parameters may make the task of evaluating boron compounds less formidable and also provide the necessary compound factors to perform accurate treatment planning for BNCT.

Calculation of Dose Distributions and Treatment Planning

Accurate methods of measuring the thermal neutron and fast neutron fluence distributions have been well developed, and neutron and gamma dose distributions in tissue equivalent phantoms have been measured and documented (Raaijmakers et al. 1995; Harling et al. 1996b). With development of techniques for determining the microdosimetric parameters for each boron compound, the boron dose may then be calculated from measured gross boron uptakes by different standard methods that have been described in the literature. These include prompt gamma spectroscopy (Kobayashi and Kanda 1983), alpha track autoradiography (Wollard et al. 1992), and atomic emission spectroscopy (Barth et al. 1991). Figure 8 shows representative central axis total RBE-dose distributions with depth inside a head for a single neutron field treatment setup for a configuration of the BMRR epithermal neutron beam. The distributions were calculated assuming a uniform normal tissue boron concentration of 6 μg/g tissue and with different uniform tumor boron concentrations for a 4-cm-deep tumor.

Monte Carlo based treatment planning techniques have been developed and are in use for treatment planning in BNCT (Wheeler and Nigg 1992; Zamenhof et al. 1990; Reinstein et al. 1993). These methods have been validated with measurements and with other calculation methods (Nigg et al. 1992). Monte Carlo based treatment planning methods suffer from the limitation that they are time consuming. Some faster calculation methods are being developed to facilitate faster treatment planning for routine BNCT use (Niemkiewicz et al. 1992, 1994).

Problems Specific to BNCT

The main attraction of BNCT, namely the ability to selectively deliver high-LET radiation to tumor, remains the biggest unknown and the greatest challenge. Since the range of the alpha particle is limited to approximately one cell diameter, cells not loaded with sufficient amounts of boron will only receive radiation doses similar to those received unavoidably by normal tissues. Cells are most resistant to conventional low-LET radiation if they are noncycling, metabolically inactive, or hypoxic; for the same reason, these same cells may prove most difficult to load with sufficient amounts of boron. A variety of strategies may have to be employed to overcome this problem, e.g., prolonged continuous drug infusions, fractionated infusion and treatment, compound combinations with different mechanisms of uptake, and BBB disruptions.

In conventional radiation therapy, fractionation is the most effective way to simultaneously enhance the effects on the tumor and to protect normal tissues. This is classically described by the 4 R's of radiation biology. Fractionation hardly moderates (if at all) the effects of high-LET radiation. In BNCT, the issue is much more complicated because of the mix of high- and low-LET radiation encountered. A further concern is the adequate delivery of boron to all parts of the tumor, as described above. Because of the highly selective delivery of boron to tumor cells, fractionation affects normal tissues and tumor differently. Since the low-LET components in adventitious radiations for epithermal neutron beams are high, fractionated dose delivery may allow the delivery of higher doses or add an additional factor of safety to normal tissues. The most important gain from fractionation is thought to be due to retargeting of the boron compound.

In order to achieve adequate tumor control, it is almost certainly necessary to use doses of a magnitude that will bring the dose to normal tissues close to tolerance levels (adventitious radiations include fast neutrons, protons from nitrogen capture reactions, gammas, boron from normal tissue uptake). The determination of a tolerance dose (TD) in BNCT is extraordinarily difficult, due to the complex mix of high- and low-LET radiations involved, the constituents of which change rapidly with depth in tissue at different rates (Gupta et al. 1994). At the same time, the possibility that a boron compound may be taken up more avidly by critical regions of the brain favoring a certain metabolic pathway, must be considered for every compound.

Neutron Capture Enhanced Fast Neutron Therapy

High-energy neutron beams produced by hospital-based accelerators are currently being used to treat cancer with fast neutrons. Neutron-nuclear reactions in tissue thermalize fast neutrons, and it is possible to use this thermalized component to initiate the neutron capture reaction and specifically enhance the radiation dose to tumor cells. Using BNCT reactions to selectively boost neutron effects in tumor cells would considerably enhance the therapeutic ratio of fast neutrons and avoid the problems of poor depth dose, poor collimation, and insufficient skin sparing.

The thermalized component in a fast neutron beam will be larger in beams with relatively low primary energy. Such neutron beams were installed in the 1960s and 1970s. Only a few of those machines are still in clinical use. One of them, the Essen Medical Cyclotron Facility [which produces a d(14)+Be-neutron beam with a mean energy of 5.8 MeV] is still of considerable clinical interest because of its isocentric gantry that allows optimization of dose distribution through the use of multiple field techniques. Treatment of deep seated tumors, however, is limited because of the steep in the depth dose distribution.

At this facility, multiple experiments resulted in the development of neutron capture assisted fast neutron therapy to optimize the depth dose distribution. Slowing down of the fast neutrons results in a distribution of partially thermalized neutrons in phantom. This distribution has been computed in a Monte Carlo model that includes a deep seated tumor (target volume), and it has been experimentally verified using Cd-covered and bare gold foils to measure the total thermal and epithermal neutron fluence rates. Distribution of the thermal neutron fluence rate in phantom depends on physical factors such as boron concentration, field size of the neutron beam (Fig. 10), beam energy (Fig. 11) and phantom geometry. Analogue Monte Carlo simulation was developed to analyze the influence of various factors on distribution of the thermal neutron fluence rate and to calculate the energy dose resulting form the ^{10}B neutron capture reaction. When a non-boronated water phantom was irradiated to a total dose of 1 Gy at a depth of 6 cm, the thermal neutron fluence was calculated to be $2 \cdot 10^{10}$ cm^{-2}. Loading the target volume with 100 ppm ^{10}B yielded a local absorbed dose enhancement of 15% at the same depth (Fig. 12).

To investigate the dependence of this enhancement on neutron beam energy, thermal neutron fluence rate and dose enhanced by ^{10}B were calculated in two other fast neutron treatment facilities: the former Chicago Cyclotron [with d(8.3)+Be-neutrons; mean energy: $E_n = 3.3$ MeV] and the Neutron Generator Hamburg [with d(0.5) + T-neutrons; $E_n = 14$ MeV]. Dose enhancements per 100 ppm ^{10}B at the same depth in phantom in these two beams were 18.5% and 7%, respectively (Fig. 11, 12).

A therapeutic gain through dose enhancement by $^{10}B(n, \alpha)^7Li$ was demonstrated in fast neutron therapy of a solid tumor model (Sauerwein 1993), in

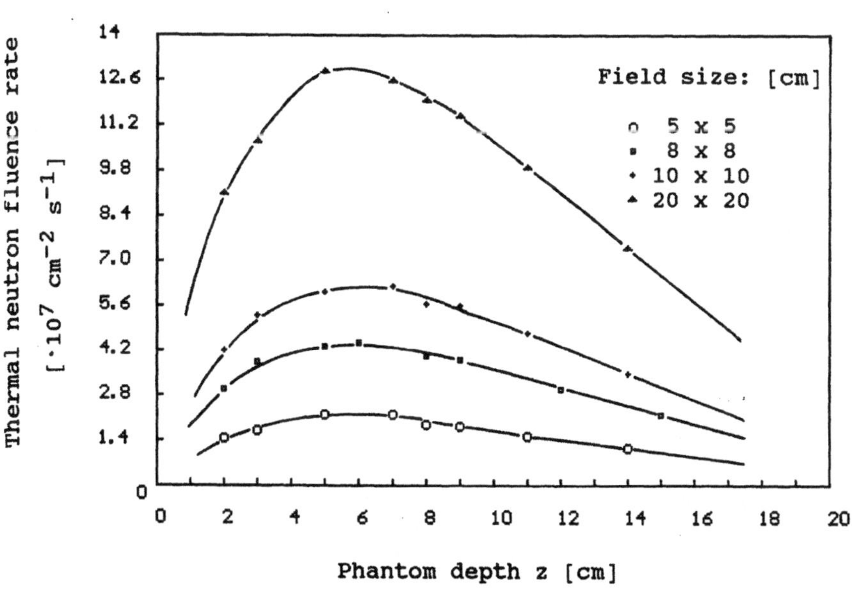

Fig. 10. Thermal neutron fluence vs depth with various field sizes

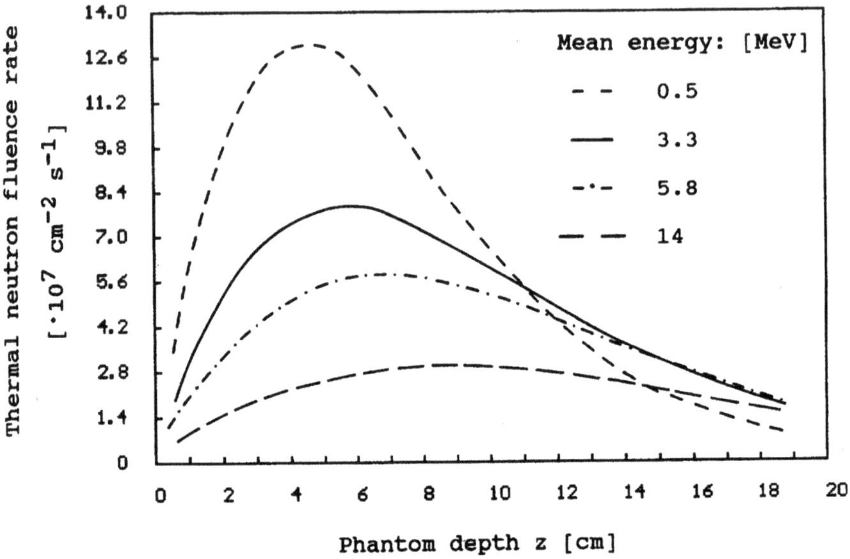

Fig. 11. Thermal neutron fluence vs depth with various energies of the primary neutron source

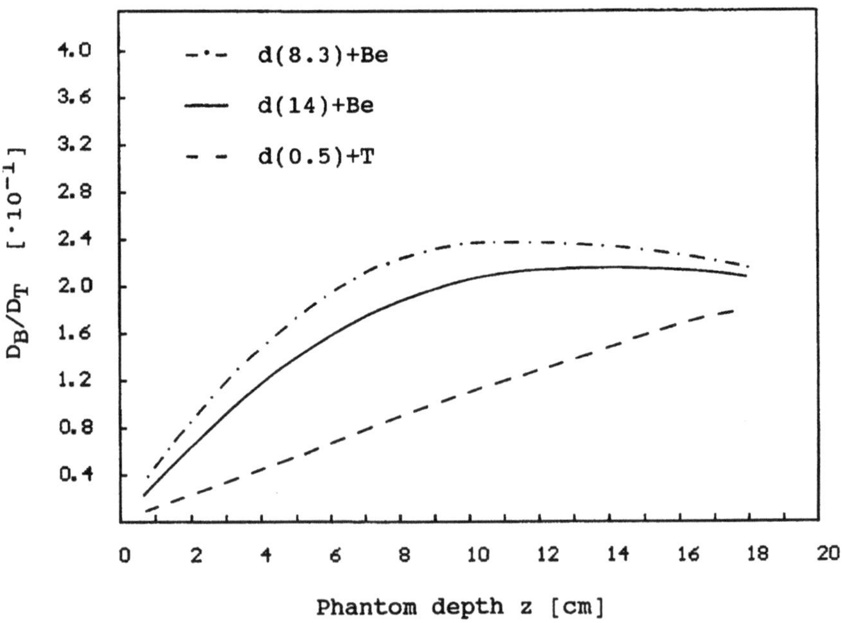

Fig. 12. "Gain factor" (CGF) as a function of depth and energy spectrum of the primary neutron source, assuming 100 ppm ^{10}B in target: $CGF = D_B \times 100/D_T$, where D_B is the boron neutron capture absorbed dose delivered by high LET alpha particles and lithium ions from neutron capture of ^{10}B by the $^{10}B(n, \alpha)^7Li$ reaction; D_T is the total absorbed dose

which the cure rate for animals receiving ^{10}B and irradiation was clearly better than that seen in the animals receiving fast neutron irradiation alone.

Clinical Trials with Thermal and Epithermal Neutrons

Early clinical trials at BNL and at MIT failed because of limitations of the neutron beams available and because boron compounds had no selectivity for tumor and inadequate clearance from blood. Using a new compound, BSH, Hatanaka reported some encouraging results from Japan, thus stimulating renewed interest in the technique worldwide. Epithermal beams were developed (Fairchild et al. 1990) to allow for the treatment of deeper seated tumors and to avoid the necessity of skull reflection in the treatment room, thus also potentially allowing for fractionated therapy in the future.

At present, clinical trials for the treatment of brain tumors have resumed at BNL and Petten, The Netherlands, and for melanomas and brain tumors at MIT. No early morbidities from either the boron compound or radiation have been observed thus far at BNL or MIT (Busse et al. 1997; Joel 1997, personal communication) or at Petten (Sauerwein 1997).

Conclusion

The developments in epithermal neutron beam technology have reached a high level of sophistication, and both reactor and non-reactor (accelerator-based) epithermal neutron sources are or can be made available if funding is available. Methods for characterizing epithermal beam performance by measurement and for verification have also been developed. Macrodosimetry tools for routine treatment planning for patients have been developed. The compounds in clinical use today are old, but ongoing clinical trials may be help define the needs for future compounds. Never compounds with higher tumor specificity and persistence, as well as improved drug delivery methods, need to be developed and tested.

References

Barth RF (1998) Rat brain tumor models in experimental neuro-oncology: the 9L, C6, T9, F98 RG-2(D74), RT-2 and CNS-1 gliomas. J Neurol-oncol 36:91–102

Barth RF, Soloway AH, Fairchild RG (1990) Boron neutron capture therapy of cancer. Cancer Res 50:1061–1070

Barth RF, Adams DM, Soloway AH, Mechetner EB, Alam F, Anisuzzaman AK (1991) Determination of boron in tissues and cells using direct-current plasma atomic emission spectroscopy. Anal Chem 63(9):890–893

Barth RF, Soloway AH, Brugger RM (1996) Boron neutron capture therapy of brain tumors: past history, current status, and future potential (Review) Cancer Invest 14(6):534–550

Barth RF, Yang W, Rotaru J, Moeschberger ML, Joel DD, Nawrocky MM, Goodman JH, Soloway AH (1997a) BNCT of brain tumors: enhanced survival following intracarotid injec-

tion of either sodium borocaptate or boronophenylalanine with or without blood-brain barrier disruption. Cancer Res 57:1129–1136
Barth RF, Yang W, Rotaru JH et al (1997b) BNCT of brain tumors: enhanced survival and cure following blood brain barrier-distribution and intracarotid injection of sodium borocaptate and boronophenylalanine. Radiother Oncol 43:515
Beach JL, Schroy CB, Ashtari M, Maruyama Y (1990) Boron neutron capture enhancement of Cf-252 brachytherapy. Int J Radiat Oncol Biol Phys 18:1421–1427
Blackburn BW, Klinkowstein RE, Yanch JC, Song H, Howard W (1997) Development of a high-power, water-cooled beryllium target for the production of neutrons in a high-current tandem accelerator. In: Duggon JL; Morgan IL (eds) Appliction of accelerators in research and industry. Proceedings of 14th international conference, Denton Tex, part 2. American Institute of Physics, Woodbury, NY, pp 1293–1296
Blue JW, Roberts WK, Blue TE, Gahbauer RA, Vincent JS (1985) A study of low energy proton accelerators for neutron capture therapy. In: Hatanaka H (ed) Neutron capture therapy. Proceedings of 2nd international symposium on neutron capture therapy. Nishimura Co. Ltd, Nigata, pp 147–158
Blue TE, Gupta N, Wollard JE (1993) A calculation of the energy dependence of the RBE of neutrons. Phys Med Biol 38:1693–1712
Blue TE, Wollard JE, Gupta N, Gahbauer RA (1995) Beam design and evaluation for BNCT. In: Proceedings of 1st international workshop on accelerator-based neutron sources for boron neutron capture therapy, Jackson, Wyo. 11–14 Sept 1994. Idaho National Engineering Laboratory, pp 197–212 (INEL report CONF-940976)
Buchholz TA, Laramore GE, Wootton P, Livesey JC, Wilbur DS, Risler R, Phillips M, Jacky J, Griffin TW (1994) Enhancement of fast neutron beams with boron neutron capture therapy. A mechanism for achieving a selective, concomitant tumor boost. Acta Oncol 33(3):307–313
Busse PM, Zamenhof R, Madoc-Jones H, Solares G, Kiger S, Riley K, Chuang C, Rogers G, Harling O (1997) Clinical follow-up of patients with melanoma of the extremity treated in a phase I boron neutron capture therapy protocol. In: Larsson B, Crawford J, Weinreich R (eds) Advances in neutron capture therapy, vol 1: medicine and physics. Elsevier, Amsterdam, pp 60–64
Capala J, Barth RF, Bendayan M, Lauzon M, Adams DM, Soloway AH, Fenstermaker RA, Carlsson J (1996) Boronated epidermal growth factor as a potential targeting agent for boron neutron capture therapy of brain tumors. Bioconjugate Chem 7(1):7–15
Christensen RN, Guo P, Blue TE, Button GL, Bahney RH, Worthy K (1991) Critical heat flux experiment for BNCT heat-removal system design. Trans Am Nucl Soc 64:647
Clendenon NR, Barth RF, Gordon WA, Goodman JH, Alam F, Staubus AE, Boesel CP, Yates AJ, Moeschberger ML, Fairchild RG, Kalef-Ezra J (1990) Boron neutron capture therapy of a rat glioma. Neurosurgery 26(1):47–55
Dahlborg SA, Henner WD, Crossen JR et al (1996) Non-AIDS primary CNS lymphomia: first example of a durable response in a primary brain tumor using enhanced chemotherapy delivery without cognitive loss and without radiotherapy. Cancer J Sci Am 2:166–174
Deutsch OL, Murray BW (1975) Monte Carlo dosimetry calculations for boron neutron capture therapy in the treatment of brain tumors. Nucl Tech 26:320–339
Faillot T, Magdelenat H, Mady E, Stasiecki P, Fohann D, Gropp P, Poisson M, Delttre JY (1996) A phase I study of anti-epidermal growth factor receptor monoclonal antibody for the treatment of malignant gliomas. Neurosurgery 39:478–483
Fairchild RG, Saraf SK, Kalef-Ezra J, Laster BH (1990) Comparison of measured parameters from a 24-kev and a broad spectrum epithermal neutron beam for neutron capture therapy: an identification of consequential parameters. Med Phys 17:1045–1052
Feakes DA, Shelly KJ, Hawthorne MF et al (1993) Liposomal delivery of boron to murine tumors for boron neutron capture therapy. In: Soloway AH, Barth RF, Carpenter DE (eds) Advances in neutron capture therapy. Plenum, New York, pp 395–398
Fenstermacher CD, Cowles AL (1977) Theoretic limitation of intracarotid infusion in brain tumor chemotherapy. Cancer Treat Rep (61):519–526

Gabel D, Foster S, Fairchild RG (1987) The Monte Carlo simulation of the biological effect of the 10 B(n, alpha) 7 Li reaction in cells and tissue and its implication for boron neutron capture therapy. Radiat Res 111(1):14–25
Gahbauer R, Goodman J, Blue T (1989) Some thoughts on tolerance, dose, and fractionation in boron neutron capture therapy. Basic Life Sci 50:81–85
Gahbauer RA, Fairchild RG, Goodman JH, Blue TE (1992) Can relative biological effectiveness be used for treatment planning in boron neutron capture therapy? In: Breit A (ed) Tumor response monitoring and treatment planning. Springer, Berlin Heidelberg New York
Gahbauer R, Gupta N, Blue T, Grecula J, Goodman J, Wambersie A (1997) Fractionation in BNCT: what are the issues? In: Larsson B, Crawford J, Weinreich R (eds) Advanced in neutron capture therapy, vol 2. Elsevier, Amsterdam, pp 683–686
Grussel E, Conde H, Larsson B, Ronnqvist T, Sornsuntisook O, Crawford J, Reist H, Dahl B, Sjostrand NG, Russel G (1990) The possible use of a spallation neutron source for neutron capture therapy with epithermal neutrons. Basic Life Sci 54:249–258
Gumerlock MK, Neuwelt EA (1993) Chemotherapy of brain tumors: innovative approaches. In: Morontz RG, Walsh JW, Dekker M (eds) Handbood of brain tumors. Plenum, New York, pp 763–778
Gupta N, Gahbauer RA, Blue TE, Wambersie A (1994) Dose prescription in boron neutron capture therapy. Int J Radiat Oncol Biol Phys 28(5):1157–1166
Hamm RW, Shubaly MR (1995) A pre-clinical radio-frequency quadrupole linac for boron neutron capture therapy measurements. In: Proceedings of 1st international workshop of accelerator-based neutron sources for boron neutron capture therapy, Jackson, Wyo., 11–14 Sept 1994. Idaho National Engineering Laboratory, pp 55–56 (INEL report CONF 940976)
Hariharan JR, Wyzlic IM, Soloway AH (1995) Synthesis of novel boron-containing polyamine agents for DNA targeting in neutron capture therapy. Polyhedron 14:823–825
Harker YD, Harmon JF (1997) Accelerator neutron sources for NCT using near threshold charged particle reactions. In: International conference on neutrons and their applications, Greece, 1996 (abstract)
Harling OK, Kiger S, Redmond EL II (1996a) High intensity fission converter based epithermal neutron beam for boron neutron capture therapy. In: Mishima Y (ed) Proceedings of the 6th international symposium of neutron capture therapy. Plenum, New York, pp 407–412
Harling OK, Moulin D, Chabeuf JM, Solares GR (1996b) On-line beam monitoring for boron neutron capture therapy at the MIT research reactor. In: Mishima Y (ed) Proceedings of the 6th international symposium of neutron capture therapy. Plenum, New York, pp 261–269
Hatanaka H, Nakagawa Y (1994) Clinical results of long-surviving brain tumor patients who underwent boron neutron capture therapy. Int J Radiat Oncol Biol Phys 28(5):1061–1066
Howard WB, Yanch JC, Grimes SM, Massey TN et al (1996) Measurement of the 9 Be(p,n) thick target spectrum for use in accelerator-based boron neutron capture therapy. Med Phys 23(7):1233–1235
Huang PC, Vafai K (1994) Internal heat transfer augmentation in a parallel plate channel using an alternate set of porous cavity-block obstacles. Numerical Heat Transfer J 25(PtA):519–540
ICRU (1990) Radiation quantities and units. (ICRU report 33, ISBN 0-913394-270) ICRU, Washington
Jones JL, Yoon WY (1992) Feasibility study of the application of a linear electron accelerator to BNCT. In: Proceedings of 12th international conference on the application of accelerators in research and industry. University of North Texas, Denton, Tex (abstract)
Kahl SB, Koo MS (1992) Synthesis and properties of tetrakis-carborane-carboxylate esters of 2,4-bis(amb-dihydroxyethyl) deuteroporphyrin IX. In: Allen BJ, Moore DE, Harrington BV (eds) Progress in neutron capture therapy for cancer. Plenum, New York, pp 223–226

Kahl SB, Pate DW, Waunschel LA (1993) Low density lipoprotein reconstitutions with alkyl and aryl carboranes. In: Soloway AH, Barth RF, Carpenter DE (eds) Advances in neutron capture therapy. Plenum, New York, pp 399–402

Kalend AM, Bloomer WD, Epperly MW (1995) Dosimetric consequences of 10 B(n, alpha) 7 Li reaction occuring at the cellular membrane. Int J Radiat Oncol Biol Phys 31(1):171–178

Kim J, Yanch JC, Wilson MJ (1993) Californium-based epithermal neutron beams for neutron capture therapy. In: Soloway AH, Barth RF, Carpenter DE (eds) Advances in neutron capture therapy. Plenum, New York, pp 131–134

Kobayashi T, Kanda K (1983) Microanalysis system of ppm-order 10 B concentrations in tissue for neutron capture therapy by prompt gamma-ray spectrometry. Nucl Instr Methods 204:96–103

Kobayashi T, Kanda K (1993) Analytical dose calculation in cell nucleus from spherical 10 B containing region and a concept of RBE for neutron capture therapy. In: Soloway AH, Barth RF, Carpenter DE (eds) Advances in neutron capture therapy. Plenun, New York, pp 217–220

Kobayashi T, Sakurai Y, Kanda K (1993) Study on remodelling the heavy water facility of the Kyota University reactor for neutron capture therapy from the concept of neutron energy spectrum control. In: Soloway AH, Barth RF, Carpenter DE (eds) Advances in neutron capture therapy. Plenum, New York, pp 29–32

Kwan JW, Anderson OA, Reginato LL, Vella MC, Yu SS (1995) Electrostatic quadrupole DC accelerators for BNCT applications. In: Proceedings of 1st international workshop on accelerator-based neutron sources for boron neutron capture therapy, Jackson, Wyo, 11–14 Sept 1994. Idaho National Engineering Laboratory, pp 111–120 (INEL report CONF-940976)

Laramore G, Diener-West M, Griffin T, Nelson J, Griem M, Thomas FJ, Handrikson F, Griffen B, Myrianthopoulos L, Saxton J (1988) Randomized neutron dose searching study for malignant gliomas of the brain: results of an RTOG study. Int J Radiat Oncol Biol Phys 14:1093–1102

Laramore GE, Wootton P, Livesey JC, Wilbur DS, Risler R, Phillips M, Jacky J, Buchholz TA, Griffin TW, Brossard S (1994) Boron neutron capture therapy: a mechanism for achieving a concomitant tumor boost in fast neutron radiotherapy. Int J Radiat Oncol Biol Phys 28(5):1135–1142

Lemmen P, Werner B, Streiches B (1993) Ether lipids as potential boron carriers for boron neutron capture therapy: synthesis of rac-1-(9-0-carboranyl)nonyl-2-methyl-glycero-3-phosphocholine (B-Et-11-Ome). In: Soloway AH, Barth RF, Carpenter DE (eds) Advances in neutron capture therapy. Plenum, New York, pp 297–300

Lieberman DM, Laske DW, Morrison PF, Bankiewicz KS, Oldfield EH (1995) Convection-enhanced distribution of large molecules in gray matter during interstitial drug infusion. J Neurosurg 82(6):1021–1029

Liu HB, Brugger RM, Rorer DC, Tichler PR, Hu JP (1994) Design of a high-flux epithermal neutron beam using 235 U fission plates at the Brookhaven Medical Research Reactor. Med Phys 21(10):1627–1631

Liu L, Barth RF, Adams DM, Soloway AH, Reisfeld RA (1995) Bispecific antibodies as targeting agents for boron neutron capture therapy of brain tumors. J Hematother 4(5):477–483

Locher GL (1936) Biological effects and therapeutic possibilities of neutrons. Am J Roentgenol 36:1–13

Matsumoto K (1993) Evaluation of depth-dose distributions for gadolinium neutron capture therapy. In: Soloway AH, Barth RF, Carpenter DE (eds) Advances in neutron capture therapy. Plenum, New York, pp 235–240

Matsukado K, Inamura T, Nakano S, Fukui M, Bartus RT, Black KL (1996) Enhanced tumor uptake of carboplatin and survival in glioma-bearing rats by intracarotid infusion of bradykinin analog, RMP-7. Neurosurgery 39(1):125–133

Miley GH, Javedani JB, Gu YB et al (1995) The IEC: a novel source for boron neutron capture therapy. In: Proceedings of 1st International workshop on accelerator-based neutron sources for boron neutron capture therapy, Jackson, Ayo, 11–14 Sept 1994. Idaho National Engineering Labortory, pp 111–120 (INEL report CONF-940976)

Mill AJ, Morgan GR, Newman SM (1994) Cell survival measurements in an argon, aluminium and sulphur filtered neutron beam: a comparison with 24 keV neutrons and relevance to boron neutron capture therapy. Br J Radiol 67(802):1008–1016

Moss RL (1990) Progress towards boron neutron capture therapy at the High Flux Reactor Petten. Basic Life Sci 54:169–183

Neuwelt EA, Frenkel EP (1989) The challenge of the blood-brain barrier. In: Neuwelt EA (ed) Implications of the blood-brain barrier and its manipulations. Plenum Press, New York, pp 1–24

Niemkiewicz J, Gupta N, Blue TE (1992) Point kernel 1 H[n,g]2 H dose calculation in BNCT. Trans Am Nucl Soc 65:155–157

Niemkiewicz J, Blue TE, Gupta N (1994) A calculation of neutron flux distributions in BNCT using removal-diffusion theory. Trans Am Nucl Soc 70:13–15

Nigg DW, Randolph PD, Wheeler FJ (1991) Demonstration of three-dimensional deterministic radiation transport theory dose distribution analysis for boron neutron capture therapy. Med Phys 18(1):43–53

Nigg DW, Mitchell HE, Harker YD, Yoon WY, Jones JL, Harmon JF (1995) Epithermal photoneutron source studies for BNCT. In: Proceedings of 1st international workshop on accelerator-based neutron sources for boron neutron capture therapy. Jackson, Wyo, 11–14 Sept 1994. Idaho National Engineering Laboratory, pp 373–386 (INEL report CONF-940976)

Pardridge WM (1995) Transport of small molecules through the blood-brain barrier: biology and methodology. Adv Drug Delivery Rev 15:5–36

Pardridge WM (1997) Drug delivery to the brain. J Cereb Blood Flow Metab 17:713–731

Raaijmakers CP, Konijnenberg MW, Verhagen HW, Mijnheer BJ (1995) Determination of dose components in phantoms irradiated with an epithermal neutron beam for boron neutron capture therapy. Med Phys 22(3):321–329

Rapoport SI, Fredericks WR, Ohno K, Pettigrew KD (1980) Quantitative aspects of reversible osmotic opening of the blood-brain barrier. Amer J Physiol 238:R421–R431

Reinstein LE, Ramsay EB, Gajewski J, Rammamoorthy S, Meek AG (1993) SBNCT-PLAN: a 3-dimensional treatment planning system for boron neutron capture therapy. In: Soloway AH, Barth RF, Carpenter DE (eds) Advances in neutron capture therapy. Plenum, New York, pp 171–175

Riva P, Arista A, Tison V, Sturiale C, Franceschi G, Spinelli A, Riva N, Casi M, Moscatelli G, Frattarelli M (1994) Intralesional radioimmunotherapy of malignant gliomas. An effective treatment in recurrent tumors. Cancer 733 Suppl 3:1076–1082

Rong FG, Soloway AH (1994) Synthesis of 5-tethered carborane-containing pyrimidine nucleosides as potential agents for DNA incorporation. Nucleosides Nucleotides 13:2021–2034

Rydin RA, Deutsch OL, Murray BW (1976) The effect of geometry on capillary wall dose for boron neutron capture therapy. Phys Med Biol 21:134–138

Salazar OM, Rubin P, Feldstein ML, Pizzutiello R (1979) High dose radiation therapy in the treatment of malignant gliomas: final report. Int J Radiat Oncol Biol Phys 5(10):1733–1740

Sauerwein W (1993) Neutroneneinfangreaktionen zur Optimierung der Strahlentherapie mit schnellen Neutronen. Habilitationsschrift, University of Essen, Germany

Sauerwein W (1997) The clinical project of HFR-Petten – a status report. In: Larsson B, Crawford J, Weinreich R (eds) Advances in neutron capture therapy, vol I, Medicine and physics. Elsevier, Amsterdam, pp 77–82

Sauerwein W, Ziegler W, Olthoff K, Streffer C, Rassow J, Sack H (1989) Neutron capture therapy using fast neutron beam: clinical considerations and physical aspects. Strahlenther Onkol 165(2/3):208–210

Scobie M, Threadgill MD (1992) Synthesis of carborane-containing nitroimidazole compounds via mild 1.3-dipolar cycloaddition. J Chem Soc Chem Commun 13:939–941

Shapiro WR, Voorhies RM, Hiesiger EM, Sher PB, Basler GA, Lipschutz LE (1988) Pharmacokinetics of tumor cell exposure to [^{14}C] methotrexate after intracarotid administration without and with hyperosomotic opening of the blood brain and blood-tumor barriers in rat brain tumours: a quantitative autoradiographic study. Cancer Res 48(3):694–701

Shefer RE, Klinkowstein RE, Yanch JC, Brownell GL (1992) An epithermal neutron source for BNCT using a tandem cascade accelerator. In: Allen BN, Moore DE, Harrington BA (eds) Progess in neutron capture therapy for cancer. Plenum, New York, pp 119–122

Shefer RE, Klinkowstein RE, Yanch JC, Howard WB (1995) Tandem electrostatic accelerators for BNCT. In: Proceedings of 1st international workshop on accelerator-based neutron sources for boron neutron capture therapy, Jackson, Wyo, 11–14 Sept 1994. Idaho National Engineering Laboratory, pp 89–98 (INEL report CONF-940976)

Slatkin DN (1991) A history of boron neutron capture therapy of brain tumours. Postulation of a brain radiation dose tolerance limit. Brain 114(4):1609–1629

Solares GR, Zamenhof RG (1995) A novel approach to the microdosimetry of neutron capture therapy. I. High-resolution quantitative autoradiography applied to microdosimetry in neutron capture therapy. Radiat Res 144(1):50–58

Spielvogel BF, Soad A, Powell W, Tomasz J, Porter K, Shaw Br (1993) Chemical and enzymatic incorporation of boron into DNA. In: Soloway AH, Barth RF, Carpenter DE (eds) Advances in neutron capture therapy. Plenum, New York, pp 389–393

Sweet W (1983) Practical problems of the past in the use of boron-slow-neutron capture therapy in the treatment of glioblastoma multiforme. In: Brownell G, Fairchild RG (eds) Procedings of the 1st international symposium on neutron capture therapy. Brookhaven National Laboratory, p 376 (BNL-51730)

Swenson DA (1995) A new linac structure for the BNCT application. In: Proceedings of 1st international workshop on accelerator-based neutron sources for boron neutron capture therapy, Jackson, Wyo, 11–14 Sept 1994. National Engineering Laboratory, pp 121–128 (report CONF-940976)

Tamargo RJ, Langer R, Brem H (1994) Interstitial drug delivery to the central nervous system using controlled release polymers: chemotherapy for brain tumors. Methods Neurosci 21:135–149

Tjarks W, Malmquist J, Gedda L, Sjoberg S, Carlsson J (1994) Synthesis and initial biological evaluation of carborane-containing phenanthrifinum derivatives. In: Mishima Y (ed) Cancer neutron capture therapy. Proceedings of 6th international symposium on neutron capture therapy for cancer. Plenum Press, New York, p 121

Vafai K, Wang W (1992) Analysis of flow and heat transfer characteristics of an asymmetrical flat plate heat pipe. Int J Heat Mass Transfer 35:2087–2099

Walker MD, Alexander EJ, Hunt WE, MacCarty CS, Mahaley MS Jr , Mealey JJ, Norrell HA, Owens G, Ransohoff J, Wilson CB, Gehan EA, Strike TA (1978) Evaluation of BCNU and/or radiotherapy in the treatment of anaplastic gliomas. A cooperative clinical trial. J Neurosurg 49(3):333–343

Walter KA, Tamargo RJ, Olivi A, Burger PC, Brem H (1995) Intratumoral chemotherapy. Neurosurgery 37(6):1128–1145

Wang CK, Moore BR (1994) Thick beryllium target as an epithermal neutron source for neutron capture therapy. Med Phys 21(10):1633–1638

Wang CK, Blue TE, Blue JW (1989) A design study of an accelerator-based neutron irradiation facility for boron neutron capture therapy. Strahlenther Onkol 165:75–78

Wangler TP, Stovall JE, Bhatia TS, Wang CK, Blue TE, Gahbauer RA (1989) Conceptual design of an RFQ accelerator-based neutron source for boron neutron capture therapy. In: 1989 particle accelerator conference, Chicago, Ill, 20–23 March. Los Alamos National Laboratory (Article LA-UR-89-912)

Waterman FM, Kuchnir FT, Skagg LS, Bewley DK, Page BC, Attix FH (1978) The use of B-10 to enhance the tumor dose in fast neutron therapy. Phys Med Biol 23:592–602

Wheeler FJ (1992) Microdosimetric calculations for boron-drug compound factors for boron neutron capture therapy. Trans Am Nucl Soc 65:147

Wheeler FJ, Nigg DW (1992) Three dimensional radiation dose distribution analysis for boron neutron capture therapy. Nucl Sci Eng 110:16–31

Whiters HR, Peters LJ (1980) Biological aspects of radiation therapy. In: Fletcher GH (ed) Textbook of radiotherapy, 3rd edn. Lea and Febiger, Philadelphia
Wollard JE, Blue TE, Curran JF, Dobelbower MC, Busby HR (1992) An alpha auradiographic technique for spatial quantification of 10B concentrations in tissue. Nucl Sci Eng 110:96–103
Wuu CS, Amols HI, Kliauga P, Reinstein LE, Saraf S (1992) Microdosimetry for boron neutron capture therapy. Radiat Res 130(3):355–359
Wyzlic IM, Tjarks W, Soloway AH, Anisuzzaman AK, Rong FG, Barth RF (1994) Strategies for the design and synthesis of boronated nucleic acid and protein components as potential delivery agents for neutron capture therapy. Int J Radiat Oncol Biol Phys 28(5):1203–1213
Yang W, Barth RF, Carpenter DE, Moeschberger ML, Goodman JH (1996) Enhacned delivery of boronophenylalanine for neutron capture therapy by means of intracarotid injection and blood-brain barrier disruption. Neurosurgery 38(5):985–992
Zamenhof RG, Clement SD, Harling OK, Brenner JF, Wazer DE, Madoc-Jones H, Yanch JC (1990) Monte Carlo based dosimetry and treatment planning for neutron capture therapy of brain tumors? Basic Life Sci 54:283–305

Subject Index

MIX
Papier aus verantwortungsvollen Quellen
Paper from responsible sources
FSC® C105338

If you have any concerns about our products,
you can contact us on
ProductSafety@springernature.com

In case Publisher is established outside the EU,
the EU authorized representative is:
Springer Nature Customer Service Center GmbH
Europaplatz 3, 69115 Heidelberg, Germany

Printed by Libri Plureos GmbH
in Hamburg, Germany